Palgrave Studies in Green Criminology

Series Editors
Angus Nurse
Department of Criminology and Sociology
Middlesex University
London, UK

Rob White
School of Social Sciences
University of Tasmania
Hobart, TAS, Australia

Melissa Jarrell
Department of Social Sciences
Texas A&M University
Corpus Christi, TX, USA

Criminologists have increasingly become involved and interested in environmental issues to the extent that the term Green Criminology is now recognised as a distinct subgenre of criminology. Within this unique area of scholarly activity, researchers consider not just harms to the environment, but also the links between green crimes and other forms of crime, including organised crime's movement into the illegal trade in wildlife or the links between domestic animal abuse and spousal abuse and more serious forms of offending such as serial killing. This series will provide a forum for new works and new ideas in green criminology for both academics and practitioners working in the field, with two primary aims: to provide contemporary theoretical and practice-based analysis of green criminology and environmental issues relating to the development of and enforcement of environmental laws, environmental criminality, policy relating to environmental harms and harms committed against non-human animals and situating environmental harms within the context of wider social harms; and to explore and debate new contemporary issues in green criminology including ecological, environmental and species justice concerns and the better integration of a green criminological approach within mainstream criminal justice. The series will reflect the range and depth of high-quality research and scholarship in this burgeoning area, combining contributions from established scholars wishing to explore new topics and recent entrants who are breaking new ground.

More information about this series at
http://www.palgrave.com/series/14622

Jennifer L. Schally

Legitimizing Corporate Harm

The Discourse of Contemporary Agribusiness

palgrave
macmillan

Jennifer L. Schally
Penn State Harrisburg
Middletown, PA, USA

Palgrave Studies in Green Criminology
ISBN 978-3-319-67878-8   ISBN 978-3-319-67879-5   (eBook)
DOI 10.1007/978-3-319-67879-5

Library of Congress Control Number: 2017956149

Cover illustration: Pattern adapted from an Indian cotton print produced in the 19th century

Printed on acid-free paper

This Palgrave Macmillan imprint is published by Springer Nature
The registered company is Springer International Publishing AG
The registered company address is: Gewerbestrasse 11, 6330 Cham, Switzerland

In Memory of Anne C. Johnson

PREFACE

I want to begin by noting that I think it is important for researchers to locate themselves within their research in order to be transparent about their potential biases. Reflexivity is essential to qualitative social research because "the researcher is the primary 'instrument' of data collection and analysis" (Watt 2007: 82). I agree with the idea that no research is value-free. Although the proper and systematic application of certain techniques may increase the rigor of one's work, research interests and questions do not just magically come into being; rather, we begin with ourselves and our own understandings of the world. Indeed, data analysis reflects the theoretical, epistemological, and ontological assumptions of the researcher (Mauthner and Doucet 2003).

For as long as I can remember, I have cared deeply for animals. My early aspiration was to become a veterinarian, and it was not uncommon for me to feed and bring home strays in my youth. Later in my life, I participated in lobby days for the HSUS, volunteered for a cat rescue group, and donated money to various animal welfare groups, including HSUS, Best Friends Animal Society, and the World Wildlife Federation. I have also been a practicing pescatarian for about a decade or so. I decided to stop eating all meat besides fish because I wanted to reduce my participation in the harms perpetrated by industrial agriculture, namely, the harm inflicted upon animals. While I realize that the conditions of farmed fish are also harmful and that overfishing of the oceans is detrimental to the environment, I also recognize myself as an imperfect human being striving to be better; my ideal self is vegan. It is from this place of caring that I was inspired to examine how the harm of some animals is not only tolerated but is culturally legitimized in the United States. In short, I *did* enter this project with preconceived notions about industrial agriculture's role in harming animals. I do not believe we can ever enter the research scene in a neutral way, but throughout my analysis, I aspired to be as critical of myself and my methods as I was of the texts I analyzed.

Middletown, PA, USA   Jennifer L. Schally

References
Mauthner, Natasha S., and Andrea Doucet. 2003. Reflexive Accounts and Accounts of Reflexivity in Qualitative Data Analysis. *Sociology* 37: 413–431.
Watt, Diane. 2007. Becoming a Qualitative Researcher: The Value of Reflexivity. *Qualitative Report* 12: 82–101.

Outline of the Book

Chapter 1 introduces the topic of the discursive construction of corporate harm and orients the work theoretically within green criminology. This chapter also outlines the focus of the book which is to answer the question of how agribusinesses culturally legitimize their harmful practices. Chapter 2 will focus on the staggering rise of industrial agriculture and some of the general harms that result from it—particularly intensive animal farming. I will first trace the origins of industrial agriculture and emphasize some of the particular harms of raising and processing animals for food in the industrial system, looking specifically at harm to animals, harm to the environment, and harm to human health. Then, in Chap. 3, I will explore and discuss some noteworthy cases of harm perpetrated by Tyson. Cases include multiple incidents of animal harm recorded via undercover video at Tyson processing plants and contracted farms as well as multiple documented environmental harms that resulted in legal action against Tyson. The descriptions of both the general and specific harm in Chaps. 2 and 3 provide a useful context for the remainder of the book. Chapter 4 illuminates the way in which corporate harms and their legitimation are situated within a complex cultural, structural, and historical landscape. The unifying argument of this chapter is that Tyson's harm/socially responsible discourse reflects general attitudes about harm to nonhumans and corporate power, as well as weak corporate regulation. In addition, Tyson's harm/ discourse cannot be understood without also understanding the history of corporate PR or "spin," and its contemporary conduit par excellence, the corporate web page, and the particularly modern "need" for companies to project social responsibility. Chapters 5 and 6 discuss how Tyson disguises their actions toward animals and the environment by using binary discourses where Tyson aligns with ideological "good" and distances themselves from ideological "bad". Chapter 5 focuses on the distancing, whereas Chap. 6 focuses on the aligning. Chapter 7 contains the summary and conclusions of the book.

Acknowledgments

Many individuals are responsible for this work. In no particular order, I thank and acknowledge the following for their help and support with this project: Lois Presser, Michelle Brown, Sherry Cable, Elizabeth Strand, Avi Brisman, Angus Nurse, Rob White, Melissa Jarrell, Steph Carey, Josie Taylor, Zachariah Biggers, and Jason Schally.

CONTENTS

LIST OF ABBREVIATIONS

AFO	Animal Feeding Operation
ALDF	Animal Legal Defense Fund
ASPCA	American Society for the Prevention of Cruelty to Animals
AVMA	American Veterinary Medical Association
CAFO	Concentrated Animal Feeding Operation
CSR	Corporate Social Responsibility
EPA	Environmental Protection Agency
EU	European Union
FBI	Federal Bureau of Investigation
FDA	Food and Drug Administration
FSIS	Food Safety Inspection Service
FTC	Federal Trade Commission
GHG	Greenhouse Gas
HSUS	Humane Society of the Unites States
NPDES	National Pollutant Discharge Elimination System
OSHA	Occupational Safety and Health Administration
PETA	People for the Ethical Treatment of Animals
PR	Public Relations
RMP	Risk Management Plan
SEC	Securities and Exchange Commission
SEP	Supplementary Environmental Project
USDA	United States Department of Agriculture

LIST OF TABLES

CHAPTER 1

Introduction

Abstract This chapter introduces the topic of the discursive construction of corporate harm and orients the work theoretically. The chapter also outlines the focus of the book, which is to answer the question of how agribusinesses culturally legitimize their harmful practices.

Keywords Agribusiness • Animal harm • Green criminology •
Harm perspectives • Discourse

Increasingly, agricultural production in the United States is consolidated in the hands of a very few powerful corporations. For example, just four companies (Tyson, Cargill, Smithfield, and JBS) make over 85 percent of all beef that is sold in the United States (Napach 2014). Tyson Foods, based in Springdale, Arkansas, is one such corporation. In the 2016 fiscal year, Tyson, on average, killed 35 million chickens, 125,000 cows, and 415,000 pigs *per week* (Tyson Foods 2016). Tyson is emblematic of "agribusiness," or interest in the profitability of food production, and the rise of "industrial agriculture." Industrial agriculture describes agricultural production that is conducted via intensive farming practices. That is, large quantities of resources (e.g., labor, fossil fuels) and technological advances (e.g., machinery, irrigation, genetic selection) are utilized to produce the highest yields from the smallest amount of space. I will be using the terms "industrial agriculture," "industrial farming," and "intensive farming" interchangeably

© The Author(s) 2018
J.L. Schally, *Legitimizing Corporate Harm*, Palgrave Studies in Green Criminology, https://doi.org/10.1007/978-3-319-67879-5_1

throughout the book as all of these terms refer to the system of agricultural production that is currently dominant in the United States.

A concomitant of agribusiness concentration is harm, especially to nonhuman animals[1] and the ecological environment.[2] Although the agricultural industry in the United States is subject to government regulation, many of the harms that are perpetrated by the industry fall within the bounds of the law. In particular, the harm inflicted upon animals on the industrialized farm proceeds on a mass scale, much—though certainly not all—of it perfectly legal.

For this project, I use Presser's definition of harm, "trouble caused by another" (2013: 2). Applying this definition to harm to the environment, harm/trouble is anything that would threaten or reduce the ability of the ecological environment to sustain life. To apply this definition to animal harm, I also borrow from Agnew's definition of animal abuse: "any act that contributes to the pain or death of an animal or that otherwise threatens the welfare of an animal" (1998: 179). Pain and death can certainly be considered "trouble." I will also talk of "animal suffering" in this book, and by that I mean a negative emotional state caused by adverse events. Although there is some debate as to whether animals can suffer (see Dawkins 2008; Rowman 1988), I take the position of the evolutionary biologist Marc Bekoff that animals are capable of experiencing a range of emotions, including suffering (Bekoff 2010).

This book investigates the discursive construction of harm and "business-as-usual" by US agribusiness by means of a case study of corporate behemoth Tyson Foods. The overarching point of this project is to demonstrate just how such harms are normalized through dominant discourses. Specifically, utilizing critical discourse analysis, I examine the ways in which Tyson Foods culturally legitimizes harm-doing through the use of discourses on their[3] corporate website.

I chose Tyson Foods as an exemplar of big agribusiness because it is the largest US-based corporation involved in livestock production,[4] with net sales exceeding $37 billion for the 2016 fiscal year. Tyson employs more people (114,000) and operates more plants (107) than do any of its competitors (Tyson Foods 2016). Like many other chicken producers, Tyson is characterized by vertical integration—meaning they control/own most all aspects of chicken production. Tyson owns the chickens they eventually slaughter and package; they contract farmers to raise the chickens to their specifications. For the 2013 fiscal year, Tyson contracted with 5500 poultry farmers (Tyson Foods 2013a). Although the cattle and hog industries are not similarly vertically integrated, Tyson maintains partnerships with many suppliers who rely on Tyson as a purchaser of large quantities of livestock (cows and pigs). Tyson has partnerships with 7500 cattle and hog suppliers (Gazdziak 2013; Tyson Foods 2013a). Because Tyson is such a large player in meat production, it makes sense to examine their discourses for how the harms of industrial agriculture get culturally legitimized.

In order to provide some context on the point of animal harm, I now turn to a brief discussion of how this subject has been conceptualized in academic literature. Other scholars note that much of the literature on animal harm deals with the so-called link between (illegal) violence against animals, usually companion animals, and violence against humans (Beirne 1999; Taylor 2011). Early work that examined animal harm was concerned with the role of animal cruelty in childhood development. That research links harming animals in childhood to later offenses and posits animal cruelty as one of three signifiers of later sociopathic behavior, the other two being enuresis (or bed-wetting) and fire-setting (Felthous 1980; Felthous and Bernard 1979; Felthous and Kellert 1987; Hellman and Blackman 1966; Wax and Haddox 1974). Other research investigates the co-occurrence of animal abuse with other family violence such as spousal or child abuse (Arkow 1997; Ascione 1997; Deviney et al. 1983). More recent work continues to focus on the link between childhood cruelty to animals and later violent offenses against humans as well as the co-occurrence of violence in families (see, for instance, DeGue and Dilillo 2009; Flynn 2011; Merz-Perez et al. 2001; Schwartz et al. 2012; Tallichet and Hensley 2004).

Although work that examines the link between violence against animals and violence against humans is important, it is equally and perhaps more important to examine harm to animals that is widely accepted and legal. As South and colleagues note, focusing on the link and instances of where animal harm is connected to interhuman abuse "does not serve animals especially well because it ignores those sites where animal abuse occurs much more often and is socially acceptable and almost invisible" (South et al. 2013: 34). As such, this project is inspired by my desire to understand how some forms of animal harm are normalized and rendered "almost invisible" within the cultural context of the United States.

There is little contention that industrial agriculture inflicts major harm on animals, humans, and the environment. For example, animals that are raised for food within the large agribusiness model are forced to live in conditions that constrict their ability to satisfy their natural instincts and infringes upon their general well-being (Harrison 1966; ASPCA 2013). Small-scale farmers experience a "cost-price" squeeze whereby the profitability of their operation decreases due to the advanced technology and efficiency of large agribusiness; the small farmer is simply unable to compete (Richards et al. 2005). Consumers are also put at risk, as food-borne illnesses have increased along with the concentration of food production (Horrigan et al. 2002; Sivapalasingam et al. 2004). Ecosystems become severely unbalanced in the large-scale farm model where crop diversity, which is vital to keeping ecosystems in balance, is rare or nonexistent on large-scale farms (Altieri 2000).

Despite the fact that the harms of industrial agriculture are well documented, 99 percent of all farmed and slaughtered animals in the United States were raised on Concentrated Animal Feeding Operations (CAFOs) in 2012 (Farm Forward 2014). Intensive animal farming operations are classified by the Environmental Protection Agency (EPA) as either Animal Feeding Operations (AFOs) or CAFOs (EPA 2012). According to the EPA, AFOs are "agricultural operations where animals are kept and raised in confined situations" and they "congregate animals, feed, manure, dead animals, and production operations on a small land area" (EPA 2012). In order to be classified as an AFO by the EPA, a lot or facility (1) must have animals that "are, or will be stabled or confined and fed or maintained for a total of 45 days or more in any 12-month period," and (2) must not produce crops or vegetation (EPA 2012). In contrast to AFOs, CAFOs are defined according to the number of animals in an operation. In the United States, the EPA classifies CAFOs as small, medium, or large. In order to be defined as a CAFO, the operation must meet the definition for an AFO plus meet the animal population guidelines (EPA 2012). An operation consisting of 1000 cattle, 700 dairy cows, 2500 hogs over 55 pounds, or 125,000 chickens (if a liquid manure system is used, the number of chickens needed to qualify is only 30,000) would be considered a large CAFO (EPA 2012).

AFOs are classified as areas of potential point-source pollution pursuant to the Clean Water Act of 1972 (*Federal Water Pollution Control Act* 1972). As such, AFOs have been regulated in some capacity since the early 1970s with the advent of the National Pollutant Discharge Elimination System (NPDES), which sets effluent guidelines and standards for various industries with the goal of reducing water pollution. The EPA defined what was considered an AFO or CAFO for NPDES in 1976; the definitions and guidelines regulating AFOs and CAFOs were revised by the

EPA for the first time in 2003 and then, most recently, in 2008 (Hribar and Schultz 2010).

One of the driving forces behind the fact that such a large number of animals are farmed intensively is the high demand for meat. At the beginning of the twenty-first century, people in the United States were consuming more meat than ever before. In addition, having meat at nearly every meal has been the expectation of most people (United States Department of Agriculture 2003). Knowledge of the harms inflicted by industrial agriculture has grown among the general public due in part to the increased speed at which information can be disseminated. According to the website ChartsBin[5] (2013), however, data from the United Nations show that meat consumption in the United States continued to increase to just over 270 pounds per person for the year of 2007. Although that number declined to about 264 pounds in 2009, the United States had the largest per capita meat consumption of all 177 countries included in the data set (ChartsBin 2013). Globally, meat consumption is projected to reach a total of 300 billion tons per year by 2022 (Tyson Foods 2013a).

Analysis of discourse within animal industries points to the use of techniques of neutralization in company literature that facilitates harm to animals, specifically the denial of injury (Stibbe 2001; Sykes and Matza 1957). In this internal discourse, animals are reduced to the status of inanimate objects through the use of language. For example, live chickens are referred to by a cooking method (e.g., broiler or fryer), live cows are referred to as "beef," and injuries sustained by live animals are referred to as "damage" (Stibbe 2001).

In order to get—and to keep—people "on board," large agribusinesses must somehow mask or culturally legitimize their harmful actions. In this book, I argue that food corporations do these things partly through their websites, where they distance themselves from the factory farm[6] image and present themselves as good corporate citizens who are "stewards of the animals, land and environment" (Tyson Foods 2013b).

THEORETICAL ORIENTATION

This project is theoretically grounded within harm-doing perspectives on crime, sometimes referred to as zemiology (Hillyard et al. 2004; Presser 2013). More narrowly, this project is a work of green criminology (Lynch 1990; South et al. 2013), which may be considered a specific harm perspective.

By "harm perspectives," I refer to a multitude of views that "widen the rather narrow approach to harm that criminology offers" (Pemberton 2007: 27). Harm perspectives broaden the scope of critical criminology by taking into consideration acts of indifference as well as purposeful acts (Pemberton 2007). Box (1983) explains "the intent to harm someone may be less immoral (at least no more immoral) than to be indifferent as to whom is harmed . . . indifference rather than intent may well be the greater cause of avoidable human suffering" (19). These perspectives hold that by only focusing on those harms that have been proscribed by law, a wide range of harmful actions are excluded. Harm theorists further contend that it is important to examine those harms that are excluded because, by nature, the law legitimizes certain power relations, thus concealing the harm of some agents (Henry and Milovanovic 1996). Presser (2013) names three main reasons for scholars to theorize about and study harm as opposed to crime:

1. Studies of harm privilege the perspectives of victims.
2. At their core, studies of crime are studies of harm.
3. Researchers may uncover new correlates and patterns of offending.

I add to this list the fact that theorizing about and research on harm as opposed to crime take into consideration that criminal laws are a human construction and therefore ever-changing and relative to historical and cultural contexts (Curra 2014; Quinney 1970). Indeed, harm perspectives somewhat reduce the relativity of crime since ideas of what constitutes harm are more standard across time and space than are legalistic ideas of what constitutes crime.

Henry and Milovanovic (1996) suggest broadening the definition of "crime" as "the expression of some agency's energy to make a difference on others and it is the exclusion of those others who in the instant are rendered powerless to maintain or express their humanity" (1996: 116). This definition has nothing to do with laws or legality but considers instances of reduction and or repression of *humanity* as criminal. Similarly, Schwendinger and Schwendinger (1970) called for a redefinition of "crime" as *human rights* violations. These definitions are interesting from the perspective of one concerned with animal harm as they are explicitly anthropocentric. Thus, even the most capacious conceptions of "harm" within the criminological endeavor exclude nonhuman animals. In this enterprise, I am aligned with the social harm and constitutive criminology perspectives, but include nonhuman animals among those beings whose social worlds are of concern.

Green Criminology

The green paradigm within criminology, which covers a broad range of topics, extends the social harm perspectives just discussed as it moves beyond the traditional conception of crime as illegal acts committed against persons to include nonhuman victims and, in many instances, harms that have not been legislated against (White 2013). The origins of the green criminological perspective are often traced to Michael J. Lynch's work in the *Critical Criminologist* in 1990 at a time when concerns about the environment became more prominent in public discourse and policy (Potter 2010). However, as other scholars have pointed out (see Goyes and South 2017; White and Heckenberg 2014), studies of environmental harms predate the coining of the term and can be found at least as early as the 1970s. Green criminology became more prominent and established as a subfield with the publication of a special issue of *Theoretical Criminology* in May of 1998. Then, in the early 2000s, efforts were made by scholars to clarify the perspective (e.g., Halsey 2004; Lynch and Stretesky 2003). Lynch and Stretesky (2003) argue that clearly defining the term "green" is necessary for the forward movement of the green criminological perspective. They contend that certain interpretations of the term lead to a green criminology that is aligned with corporate interests, focusing only on violations of environmental regulations, while ignoring real harm that is caused by pollution, dumping, and use of resources that is done within legal limits. The authors argue that the "green" in green criminology must be redefined through a frame of environmental justice, which recognizes the detrimental effects of legitimate activities. Halsey has argued that "green" should not be used at all to describe the perspective due to the fact that "it does not adequately capture the inter-subjective, inter-generational, or inter-ecosystemic processes which combine to produce scenarios of harm" (2004: 835).[7]

Debate continues as to exactly how this perspective should be labeled, but criminologists mainly use "green criminology" (South et al. 2013). Despite the attempts to more clearly define (and name) the subfield, the framework remains quite broad and flexible, with green criminological works covering a range of topics including climate change, pollution, and the trade of endangered species (South et al. 2013). Some green criminological works have adopted a legalistic view by restricting their study of environmental harms to those acts or omissions that are expressly proscribed by law (e.g., Shover and Routhe 2005)—such studies have been referred to as "legal-procedural" (South et al. 2013). Other green criminologists adopt a "socio-legal approach," questioning both illegal and legal practices that are harmful to the environment and/or animals (South et al. 2013). This text follows the latter approach, aligning with Beirne and South's hope for green criminology to "be a harm-based discourse that addresses...environmental morality, environmental ethics, and animal rights" (2007: xiv). Within the context of Beirne and South's vision of green criminology, there is a clear space for a critical discourse analysis of Tyson Foods' website as an exemplar of how the harm perpetrated by large agribusinesses, meat producing ones in particular, gets culturally legitimized.

Particularly relevant to this project is Brisman and South's (2013, 2014) recent attempts to integrate green criminology with cultural criminology. Cultural criminology pays particular attention to how crime and crime control are situated culturally. That is to say, cultural criminologists are interested in how understandings of deviant and/or criminal behaviors are constructed through cultural processes including the use of symbols and interaction with mass media. This perspective has traditionally shown that the so-defined deviant/criminal actions of individuals and groups, such as various youth subcultures, are reactions to oppressive conditions of late modernity. Moreover, cultural criminology is about power and how power is, specifically, conjured within the culture of the late-modern era. The intersection of cultural and green criminology makes sense because it is in this time of late modernity that the so-called green issues are at the forefront of global culture, with the Internet representing the foremost place where cultural symbolism is produced and reproduced.

This project examines discursive foundations of harm to animals, and the environment in the age of the Internet. Brisman and South (2013, 2014) note the absence of green criminological work on the construction of harm in various "mediated representations or constructions of environmental crimes and harms" outlets (2013: 11). In particular, the role of discourse in producing and reproducing harms has received limited attention within green criminology. There are a few notable exceptions to this exclusion. For example, White (2008) describes how environmental problems, like other social problems, are socially constructed "through a combination of material and cultural factors" (2008: 32). Kalof and Taylor are concerned with the competing discourses surrounding dogfighting—discourses of those who fight dogs, the "worried public," and law enforcement (2007: 319). Riise examines one Norwegian political document to "show how language is employed by the authorities to shape human-animal relationship; how the exploitation of non-humans is being upheld through the use of conceptual power" (2012: 134). The current project seeks to add to our understanding of how language shapes harmful actions, the victims of which are not human.

DISCURSIVE FOUNDATIONS OF HARM

Social action is constituted in discourse. That is, motivations and justifications for human actions are socialized. We come to the world of discourses that preexist us but are prone to creative use and change; these shape how we view the world—which behaviors are and are not acceptable given our social locations. According to Berger and Luckmann (1967), society is both an objective and a subjective reality and these realities interact dialectically: people create the society in which they live and people are in turn shaped by society. Over time, the "objectivity of the institutional world 'thickens' and 'hardens'" such that it is thought of as not only normal, but natural (Berger and Luckmann 1967: 59).

Demonstrating just how normalized harm can become within certain discourses, Cohn (1987) writes of her experience working at a center on defense technology and arms control. Her experience began as a two-week summer workshop that culminated into a more permanent position for Cohn. At the start, she described the defense experts who presented the workshop as engaging in "dispassionate discussion of nuclear war," and she was baffled as to how they could think the way they did (Cohn 1987: 688). One of Cohn's observations included that "[n]uclear missiles are based in 'silos.' On a Trident submarine, which carries twenty-four multiple warhead nuclear missiles, crew members call the part of the submarine where the missiles are lined up in their silos ready for launching 'the Christmas tree farm.' What could be more bucolic-farms, silos, Christmas trees?" (Cohn 1987: 698). Her initial bafflement notwithstanding, in having to learn to speak the language of nuclear planners, Cohn discovered that part of what made it easy to so unemotionally talk about such a thing as nuclear war was how the language distanced speakers from having to actually acknowledge what it was they were talking about—although this was not a conscious choice:

Learning to speak the language of defense analysts is not a conscious, cold-blooded decision to ignore the effects of nuclear weapons on real live human beings, to ignore the sensory, the emotional experience, the human impact. It is simply learning a new language, but by the time you are through, the content of what you can talk about is monumentally different, as is the perspective from which you speak. (Cohn 1987: 705)

Within the new discourse she learned, nuclear war became a mundane topic—just another thing that gets talked about at work every day. Indeed, Cohn states that the more that she talked about it, the less afraid of nuclear war she became, as if the words somehow erased the reality of it.

In his book *Why War?*, Philip Smith (2005) identifies discourses that have helped to facilitate US declarations of war. He singles out for attention the opposing discourses of liberty and repression: the "Discourse of Liberty" consists of the ideas of open, trusting relationships that "involve the free and fair exchange of ideas," whereas the "Discourse of Repression" involves relationships that are characterized by deceit, suspicion, and secrecy (2005: 16). Smith notes that successfully associating enemies with the "Discourse of Repression" is a necessary precondition for war.

Other scholars have looked at discourses or logics that permit harm to animals (Presser 2013; Presser and Schally 2013; Stibbe 2001). Regarding meat-eating, Presser (2013) found that people discursively construct themselves as both powerful and powerless, and that this so-called power paradox permits the practice to continue. Individuals expressed being able to eat the flesh of animals because essentially "that's what they're for," an idea rooted in discourses of "custom, utility, and religion" (Presser 2013: 55). Individuals also expressed a helplessness regarding meat-eating—an inability to stop eating meat or to stop meat production. The discursive constructions allowed the participants to be complicit in the harm that is intensive animal farming. Where harm to humans is often precipitated by dehumanization, Presser notes that harm to animals does not require the same "reduction of target" for it is coded in the language of dehumanization that "to be nonhuman is to risk mistreatment...nonhumans cannot logically be dehumanized" (2013: 53).

In their work concerning discourses that led up to an amendment to the Tennessee state constitution, which instituted hunting and fishing as a right, Presser and Schally (2013) discovered three main logics that facilitated the passing of the amendment: the discourse of economic utility, veneration of the past, and claims of future infringement. In the discourse of economic utility, harm is legitimized because it is profitable. Discourses that venerated the past legitimized the harm of hunting and fishing by associating it with long-standing traditions that were categorized as "good." Discursive constructions of victimhood were the most prominent: those in support of the amendment claimed a need to be protected from animal rights groups, activist judges, and a legislature which might "willy-nilly" decide to outlaw hunting and fishing (Presser and Schally 2013: 179).

Stibbe's (2001) work also exposes the discursive foundations of animal harm. A critical discourse analysis of articles from meat industry trade magazines (e.g., *Poultry* and *Meat Marketing and Technology*), professional journals (e.g., veterinary and law journals), and some articles written by the meat industry to justify farming methods led Stibbe to the observation that animal harm is normalized through discourses of science which serve to naturalize the oppression of animals. Use of animals for food was presented through the lens of the biological rules of predation. Other logics Stibbe found included the casting of animals as inanimate resources and animal harm being obscured through the use of linguistic devices such as nominalization and metonymy (see Appendix for explanation of nominalization and metonymy). The current project follows Stibbe's example but broadens the scope of discourses examined by looking at a different type of medium, namely, the corporate website. Further, analysis of the corporate website places attention on discourse intended for the public, which may reveal more diverse discursive constructions than those found by Stibbe in his research on industry 'talk'.

This project is an exploration of several understudied topics in criminology in general: harm to animals, legal harm, and the role of discourse in perpetuating harm. These topics generally fall outside of the scope of traditional criminological work, which tends to focus on illegal behaviors—specifically, street crime and individual offenders. This book seeks to add to those green criminological works that consider discursive contexts of harm to animals and the environment by examining the role that discourses play in producing/perpetuating a majority of citizens who silently consent to the harms of industrial agriculture. This project further expands criminological research on animal harm by including harms that are not proscribed by law.

NOTES

1. Beirne (1999) notes that although the use of the term "nonhuman animal" serves the purpose of connoting that humans are also animals, the term is not without problems. Beirne likens the use of the term "nonhuman animals" to the use of the term "non-male human," where one person is defined negatively in terms of their relation to another. Nonetheless, the terms "animals" and "nonhuman animals" will be used interchangeably throughout the book.

2. Please note that the separation between environment, humans, and animals is merely practical. I recognize that a broad definition of environment necessarily includes people and animals. Later in the book, I discuss specific harms and these categorizations (environment, humans, and animals) facilitate that discussion.

3. I will use pronouns such as "they," "them," and "their" to refer to Tyson and other corporations throughout the book. Although these pronouns personify the corporation, referring to the corporations as "it" poses a greater issue as it discursively constructs the company as a nonagent (see Korten 1995).

4. Note that the phrases "livestock producers" and "chicken producers" will be used throughout the book, although I find these terms inherently problematic as they imply that the animals require human intervention to exist. The terms "pork producers" and "beef producers" cause slightly less concern because the terms "pork" and "beef" refer not to the animals themselves, but to the meat that comes from these animals. These are examples of metonymy, explained in Appendix.

5. ChartsBin is an online tool for creating charts and graphs. The ChartsBin website provides interactive charts and maps for several public data sets.

6. The term "factory farm" is often used by animal activists to describe intensive animal farming as it connotes the poor conditions animals experience within industrial agriculture. The term is considered offensive, inaccurate, and nontechnical by those in the industry (see McCarty 2005).

7. Note that Halsey later distanced himself from his 2004 work, stating that he is "less wedded" to the idea that "the term 'green' should be jettisoned from criminological discourse" (Halsey 2013: 107).

REFERENCES

Agnew, Robert. 1998. The Causes of Animal Abuse: A Social-Psychological Analysis. *Theoretical Criminology* 2: 177–209.

Altieri, Miguel A. 2000. Ecological Impacts of Industrial Agriculture and the Possibilities for Truly Sustainable Farming. In *Hungry for Profit: The Agribusiness Threat to Farmers, Food, and the Environment*, ed. F. Magdoff, J.B. Foster, and F.H. Buttel. New York: Monthly Review.

Arkow, Phil. 1997. Relationships Between Animal Abuse and Other Forms of Family Violence. *Protecting Children* 13: 4–9.

Ascione, Frank R. 1997. Battered Women's Report of Their Partner's and Their Children's Cruelty to Animals. *Journal of Emotional Abuse* 1: 1–12.

ASPCA. 2013. *What Is a Factory Farm?* New York: American Society for the Prevention of Cruelty to Animals. http://www.aspca.org/Fight-AnimalCruel ty/farm-animal-cruelty/what-is-afactory-farm. Accessed 18 Aug 2013.

Beirne, Piers. 1999. For a Non-Speciesist Criminology: Animal Abuse as an Object of Study. *Criminology* 37: 117–147.

Beirne, Piers, and Nigel South. 2007. Introduction: Approaching Green Criminology. In *Issues in Green Criminology*, ed. Piers Beirne and Nigel South, xiii–xxii. Portland: Willan.

Bekoff, Marc. 2010. *The Animal Manifesto: Six Reasons for Expanding Our Compassion Footprint*. Novato: New World Library.

Berger, Peter L., and Thomas Luckmann. 1967. *The Social Construction of Reality*. New York: Anchor Books.

Box, Steven. 1983. *Power, Crime, and Mystification*. London: Tavistock.

Brisman, Avi, and Nigel South. 2013. A Green-Cultural Criminology: An Exploratory Outline. *Crime Media Culture* 9: 115–135.

———. 2014. *Green Cultural Criminology: Constructions of Environmental Harm, Consumerism, and Resistance to Ecocide*. London/New York: Routledge.

ChartsBin. 2013. *Current Worldwide Annual Meat Consumption Per Capita*. http://chartsbin.com/view/bhy. Accessed 16 Feb 2014.

Cohn, Carol. 1987. Sex and Death in the Rational World of Defense Intellectuals. *Signs: Journal of Women in Culture and Society* 12: 687–718.

Curra, John. 2014. *The Relativity of Deviance*. Thousand Oaks: Sage.

Dawkins, Marian S. 2008. The Science of Animal Suffering. *Ethology* 114: 937–945.

DeGue, Sarah, and David DiLillo. 2009. Is Animal Cruelty a 'Red Flag' for Family Violence? Investigating Co-Occurring Violence Toward Children, Partners, and Pets. *Journal of Interpersonal Violence* 2: 1036–1056.

Deviney, Elizabeth, Jeffrey Dickert, and Randall Lockwood. 1983. The Care of Pets Within Child Abusing Families. *International Journal for the Study of Animal Problems* 4: 321–329.

EPA. 2012. *Regulatory Definitions of Large CAFOs, Medium CAFOs, and Small CAFOs*. Washington, DC: Environmental Protection Agency. http://www.epa.gov/npdes/pubs/sector_table.pdf. Accessed 16 Feb 2014.

Farm Forward. 2014. *Ending Factory Farming*. https://farmforward.com/ending-factory-farming/. Accessed 20 July 2016.

Federal Water Pollution Control Act. 1972. 33 U.S.C. 1251.

Felthous, Alan R. 1980. Aggression Against Cats, Dogs, and People. *Child Psychiatry and Human Development* 10: 169–177.

Felthous, Alan, and Harold Bernard. 1979. Enuresis, Fire Setting and Cruelty to Animals: The Significance of Two Thirds of This Triad. *Forensic Science* 2: 240–246.

Felthous, Alan R., and Stephen R. Kellert. 1987. Childhood Cruelty to Animals and Later Aggression Against People: A Review. *American Journal of Psychiatry* 14: 710–717.

Flynn, Clifton P. 2011. Examining the Links Between Animal Abuse and Human Violence. *Crime, Law and Social Change* 55 (5): 453–468.

Gazdziak, Sam. 2013. *The National Provisioner's Top 100*. The National Provisioner. http://www.provisioneronline.com/ext/resources/2013May/024-037-top-100. pdf. Accessed 25 Sept 2013.

Goyes, David Rodriguez, and Nigel South. 2017. Green Criminology Before 'Green Criminology': Amnesia and Absences. *Critical Criminology* 25: 165–181.

Halsey, Mark. 2004. Against 'Green' Criminology. *British Journal of Criminology* 44: 833–853.

———. 2013. Conservation Criminology and the "General Accident" of Climate Change. In *Routledge International Handbook of Green Criminology*, ed. N. South and A. Brisman, 107–119. New York: Routledge.

Harrison, Ruth. 1966. *Animal Machines: An Expose of "Factory Farming" and Its Danger to the Public*. New York: Ballantine Books.

Hellman, Daniel S., and Nathan Blackman. 1966. Enuresis, Fire Setting and Cruelty to Animals: A Triad Predictive of Adult Crime. *American Journal of Psychiatry* 122: 1431–1435.

Henry, Stuart, and Dragan Milovanovic. 1996. *Constitutive Criminology: Beyond Postmodernism*. London: Sage.

Hillyard, Paddy, Christina Pantazis, Steve Tombs, and Dave Gordon, eds. 2004. *Beyond Criminology: Taking Harm Seriously*. London: Pluto Press.

Horrigan, Leo, Robert S. Lawrence, and Polly Walker. 2002. How Sustainable Agriculture Can Address the Environmental and Human Health Harms of Industrial Agriculture. *Environmental Health Perspectives* 110 (5): 445–456.

Hribar, Carrie, and Mark Schultz. 2010. *Understanding Concentrated Animal Feeding Operations and Their Impact on Communities*. Bowling Green: National Association of Local Boards of Health. http://www.cdc.gov/nceh/ehs/docs/ understanding_cafos_nalboh.pdf. Accessed 25 May 2014.

Kalof, Linda, and Carl Taylor. 2007. The Discourse of Dog-Fighting. *Humanity and Society* 31: 319–333.

Korten, David C. 1995. *When Corporations Rule the World*. Bloomfield: Kumarian Press.

Lynch, Michael J. 1990. The Greening of Criminology: A Perspective for the 1990s. *The Critical Criminologist* 2: 1–4.

Lynch, Michael J., and Paul B. Stretesky. 2003. The Meaning of Green: Contrasting Criminological Perspectives. *Theoretical Criminology* 7: 217–238.

McCarty, Rick. 2005. *Consumers Aware of Factory Farming; Term Creates Negative Impression*. National Cattlemen's Beef Association. http://www.beefusa.org/uDocs/factoryfarming.pdf. Accessed 20 June 2014.

Merz-Perez, Linda, Kathleen M. Heide, and Ira J. Silverman. 2001. Childhood Cruelty to Animals and Subsequent Violence Against Humans. *International Journal of Offender Therapy and Comparative Criminology* 45: 556–573.

Napach, Bernice. 2014. *How 4 Companies Control Almost All the Meat You Eat*. Yahoo Finance. http://finance.yahoo.com/blogs/daily-ticker/how-four-companies-control-the-supply-and-price-of-beef--pork-and-chicken-in-the-u-s-eat-prices-224406080.html. Accessed 30 June 2014.

Pemberton, Simon. 2007. Social Harm Future(s): Exploring the Potential of the Social Harm Approach. *Crime, Law, and Social Change* 48: 27–41.

Potter, Gary. 2010. What Is Green Criminology? *Sociology Review*. http://www.greencriminology.org/monthly/WhatIsGreenCriminology.pdf. Retrieved 18 Aug 2013.

Presser, Lois. 2013. *Why We Harm*. New Brunswick: Rutgers University Press.

Presser, Lois, and Jennifer L. Schally. 2013. Institutionalizing Harm in Tennessee: The Right of the People to Hunt and Fish. *Journal of Sociology and Social Welfare* 40: 169–184.

Quinney, Richard. 1970. *The Social Reality of Crime*. Boston: Little, Brown and Company.

Richards, Carol, Geoff Lawrence, and Nigel Kelly. 2005. Beef Production and the Environment: Is It Really "Hard to Be Green When You Are in the Red?". *Rural Society* 15: 192–209.

Riise, Ingvill H. 2012. Natural Exploitation: The Shaping of the Human-Animal Relationship Through Concepts and Statements. In *Eco-Global Crimes: Contemporary Problems and Future Challenges*, ed. R. Ellefsen, R. Sollund, and G. Larsen, 133–156. Burlington: Ashgate.

Rowman, Andrew N. 1988. Animal Anxiety and Suffering. *Applied Animal Behaviour Science* 20: 135–142.

Schwartz, Rebecca, William Fremouw, Allison Schenk, and Laurie Ragatz. 2012. Psychological Profile of Male and Female Animal Abusers. *Journal of Interpersonal Violence* 27: 846–861.

Schwendinger, Herman, and Julia Schwendinger. 1970. Defenders of Order or Guardians of Human Rights? *Issues in Criminology* 5: 123–157.

Shover, Neal, and Aaron S. Routhe. 2005. Environmental Crime. *Crime & Justice* 32: 321–371.

Sivapalasingam, Sumathi, Cindy R. Friedman, Linda Cohen, and Robert V. Tauxe. 2004. Fresh Produce: A Growing Cause of Outbreaks of Foodborne Illness in the United States, 1973 Through 1997. *Journal of Food Protection* 67: 2342–2353.

Smith, Philip. 2005. *Why War? The Cultural Logic of Iraq, the Gulf War, and Suez*. Chicago: University of Chicago Press.

South, Nigel, Avi Brisman, and Piers Beirne. 2013. A Guide to Green Criminology. In *Routledge International Handbook of Green Criminology*, ed. N. South and A. Brisman, 27–42. New York: Routledge.

Stibbe, Arran. 2001. Language, Power and the Social Construction of Animals. *Society & Animals* 9: 145–161.

Sykes, Gresham M., and David Matza. 1957. Techniques of Neutralization: A Theory of Delinquency. *American Sociological Review* 22: 664–670.

Tallichet, Suzanne E., and Christopher Hensley. 2004. Exploring the Link Between Recurrent Acts of Childhood and Adolescent Animal Cruelty and Subsequent Violent Crime. *Criminal Justice Review* 29: 304–315.

Taylor, Nik. 2011. Criminology and Human-Animal Violence Research: The Contribution and the Challenge. *Critical Criminology* 19: 251–263.

Tyson Foods. 2013a. *Fiscal 2013 Fact Book*. http://ir.tyson.com/files/doc_do wnloads/Tyson%202013%20Fact%20Book.pdf. Accessed 17 Apr 2014.

———. 2013b. *Core Values*. http://www.tysonfoods.com/Our-Story/Core-Val ues.aspx. Accessed 11 Aug 2013.

———. 2016. *Facts About Tyson Foods*. http://ir.tyson.com/investorrelations/inve stor-overview/tyson-factbook/. Accessed 1 Dec 2016.

United States Department of Agriculture. 2003. *Agriculture Factbook 2001–2002*. Washington, DC: United States Department of Agriculture. http://www.usda.gov/factbook/2002factbook.pdf. Accessed 18 Aug 2013.

Wax, Douglas E., and Victor Haddox. 1974. Enuresis, Fire Setting, and Animal Cruelty: A Useful Danger Signal in Predicting Vulnerability of Adolescent Males to Assaultive Behavior. *Child Psychiatry and Human Development* 4: 151–156.

White, Rob. 2008. *Crimes Against Nature*. Portland: Willan.

———. 2013. *Environmental Harm: An Eco-Justice Perspective*. Bristol: Policy Press.

White, Rob, and Diane Heckenberg. 2014. *Green Criminology: An Introduction to the Study of Environmental Harm*. New York: Routledge.

Clifton P. Flynn, (2011) Examining the links between animal abuse and human violence. Crime, Law and Social Change 55 (5):453-468

CHAPTER 2

Industrial Agriculture and Its Harms

Abstract This chapter provides an overview of industrial agriculture and its harms, especially to animals, in order to provide a vivid and alarming context for my analysis of Tyson Foods' website. First, the origins of industrial agriculture are traced. The chapter continues with emphasis on some of the particular harms of raising and processing animals for food in the industrial system, specifically harm to animals, harm to the environment, and harm to human health.

Keywords Industrial agriculture • Intensive animal farming • Human health • Environment

Over 9 billion animals were killed for food in the United States in 2015—approximately 28,000 farm animals per person living in the United States that year (Humane Society of the United States (HSUS) 2016). These farm animals did not die without suffering. This chapter provides an overview of industrial agriculture and its harms, especially to animals, in order to provide context for my analysis of Tyson Foods' website. I will first trace the origins of industrial agriculture and emphasize some of the particular harms of raising and processing animals for food in the industrial system, looking specifically at harm to animals, harm to the environment, and harm to human health. The harms of industrial agriculture are numerous and complex. Therefore, this chapter should not be taken as an exhaustive account of

© The Author(s) 2018
J.L. Schally, *Legitimizing Corporate Harm*, Palgrave Studies in Green Criminology, https://doi.org/10.1007/978-3-319-67879-5_2

all of the harms that are associated with industrial agriculture, but rather should be looked at as a survey, intended to provide an overview.

THE HISTORY OF INDUSTRIAL AGRICULTURE IN THE UNITED STATES

The move toward industrial agricultural production in the United States can be traced to the Morrill Act of 1862 (*Land Grant College Act*, 7 U.S.C. 301 et seq. 1862 seq.), the federal legislation that instituted the Land Grant University System, and to the institution of the United States Department of Agriculture (USDA) in the same year (Florer 1968; Centner 2004; USDA 2014a). The Morrill Act was introduced by Justin Morrill, a Vermont state representative, for the purpose of distributing approximately 6 million acres of federal land across the states; each state was to use this land to establish at least one college where "agriculture and the mechanic arts" would be the primary focus (*Land Grant College Act*, 7 U.S.C. 301 et seq. 1862; Florer 1968: 459). At that time, the economy of the United States was largely agrarian, and Morrill argued that the United States could not afford to be second best in agricultural production, drawing comparisons to European competitors (Florer 1968). Keeping up with Europe in terms of agricultural production was a clear concern of the time as evidenced by President Abraham Lincoln's signing of the legislation that would create the USDA (USDA 2014a). The purpose of creating the USDA was to "acquire and to diffuse among the people of the United States useful information on subjects connected with agriculture in the most general and comprehensive sense of that word, and to procure, propagate, and distribute among the people new and valuable seeds and plants" (USDA 2014a).

The federal infrastructure and support provided by the Morrill Act and the institution of the USDA created a context in which farmers were encouraged to innovate and adopt new practices that would increase yields. At the same time, science and technology were advancing rapidly. For example, the development of farm machines started with the invention of the reaper by Cyrus McCormick in the 1840s and rapidly continued throughout the rest of the nineteenth century with the introduction of many other mechanized products such as the combine, poultry incubators, and mechanical planters (Pew Commission on Industrial Farm Animal Production 2008). All of the new inventions provided the tools that farmers needed to increase production yet reduce labor, which did occur (Centner 2004). Following this period of greater yields, farm yields leveled off in the mid-twentieth century before, once again, rapidly increasing in the 1960s due to further technological advances, such as genetic selection, chemical fertilizer, and pesticides (Pew Commission on Industrial Farm Animal Production 2008). The technological advances that led to this second increase in agricultural production are often collectively referred to as the "Green Revolution" (Pew Commission on Industrial Farm Animal Production 2008). As a result of the Green Revolution, agricultural production began to outpace the food needs of consumers, creating a surplus of crops—namely, corn and other grains. The surplus of corn and other grains led farmers to look to these cheap and abundant sources of food to feed livestock; by utilizing these sources, "large-scale animal agriculture [became] more profitable" (Pew Commission on Industrial Farm Animal Production 2008: 3). Thus, advances in crop production increased yields, which in turn allowed increased production within animal agriculture; for example, since 1960, milk production has doubled, meat production has tripled, and egg production has quadrupled (Centner 2004; Delgado 2003; Pew Commission on Industrial Farm Animal Production 2008).

Criticisms of intensive animal farming began to emerge early on during the second period of rapid change beginning in the 1960s, most notably by Ruth Harrison, who wrote about the treatment of chickens, pigs, cows, and veal calves on industrial farms in great detail in her 1966 book, *Animal Machines.* Harrison (1966) noted that all of the animals are forced to live in conditions that constrict their ability to satisfy their natural instincts and that their general well-being is not held in any regard. She called for the abolition of many common practices at that time and presented her work as an exposé of "factory farms" with the goal of reforming the industry. More than 50 years later, however, the operations on industrial animal farms have not improved very much. The industry's use of battery cages—small wire cages that are stacked on top of one another in rows for egg-laying hens—was one practice that Harrison (1966) spoke out against in her book. Although battery cages are in the process of being phased out in some states (California, Michigan, and Ohio), federal legislation to ban the practice of housing egg-laying hens in battery cages has not been successful (HSUS 2012). In fact, a majority (more than 90 percent) of egg-laying hens continue to be housed in battery cages (Friedrich 2013). The use of battery cages is just one example of harmful conditions on intensive animal farms. In the next section, I will describe battery cages in more detail as well as several other specific harms of intensive animal farming.

The Harms of Intensive Animal Farming[1]

Intensive animal farming is harmful to the animals that are raised for food under this system. It is also harmful to the environment and to human health. In this section, I will detail some of these harms, restricting my discussion to intensive animal farming practices in the United States.

Harm to Animals

The US federal Animal Welfare Act (*Animal Welfare Act* 1966), which was passed in 1966, authorizes the secretary of agriculture to create the standards for humane handling of animals by dealers, research facilities, and exhibitors, but excludes livestock, horses not used for research, rodents, and birds (Animal Legal Defense Fund 2014; *Animal Welfare Act* 1966). The Animal Welfare Act makes no provisions for farmed animals, and most states' animal cruelty laws exclude animals raised for food and legally hunted animals (Animal Legal Defense Fund 2014; *Animal Welfare Act* 1966). Federal legislation regarding the humane slaughter of animals—specifically, the Humane Methods of Slaughter Act of 1958—states that animals must be rendered unconscious via some quick and effective method prior to slaughter (*Humane Methods of Slaughter Act* 1958; USDA 2014b). The Humane Methods of Slaughter Act, however, excludes chickens (*Humane Methods of Slaughter Act* 1958; Animal Legal and Historical Center 2014). The near absence of laws governing the treatment of animals within the industrial agricultural model allows the exploitation and abuse of animals, both in terms of how animals are raised and in terms of how they are killed. As noted by Angus Nurse, author of *Animal Harm*, "the desire for profit and the lack of effective regulation create an environment where animal abuse is not only possible but becomes operationally acceptable as a means of maximizing profits" (2013: 39).

Severely crowded conditions are the basis of many of the harms that prevail on industrial farms. The concentration of animals is evidenced by the fact that in 2001, fewer than 30,000 poultry farms were producing nearly 8.5 billion chickens, 272 million turkeys, and 84 billion eggs each year (Centner 2004), and USDA data from 2012 indicated that approximately 69,000 hog farms raised around 65 million pigs (USDA 2012). Although concentration of cows (approximately 922,000 operations raising around 9.5 million) is not quite as severe as concentrations of chickens and pigs—owing to the fact that fewer cows and pigs are needed than chickens to produce comparable amounts of meat—the concentration of cows has been increasing over the past decade (USDA 2012).

Animals that are raised on industrial farms are subjected to extremely close living quarters so as to produce the greatest number of animals as is possible from the smallest amount of space. For breeding sows, this means being crammed into gestation crates, which have been the most common housing used for pregnant sows in the pork industry (HSUS 2014). Gestation crates are so small and so close together that the animal is barely able to move for her entire pregnancy. Breeding sows are moved to somewhat larger farrowing crates shortly before they give birth. The farrowing crates are spaced so that there is room to nurse piglets. Once the piglets are weaned, however, the sows are subjected to artificial insemination and shortly after returned to the gestation crates (Shapiro 2013). The breeding sows live this cycle for practically their entire lives (Shapiro 2013). The drive to produce more with less means that egg-laying hens are confined in battery cages (HSUS 2012). Battery cages are so small that hens cannot fully extend their wings, much less engage in other natural behaviors such as perching, nesting, and dust bathing (HSUS 2012).

Close quarters cause problems to animals in many other ways. For example, pigs experience skeletal deformities in their feet and legs and are subjected to tail removal (usually without anesthesia) to prevent other pigs from biting them in the overcrowded conditions (Jacques et al. 2013). Workers on industrial chicken farms often practice debeaking—chickens have their beaks cut off as chicks with no anesthesia, a practice that has been banned in Sweden since 1999—to prevent them from injuring each other in ways that would result in a financial loss (Hirsch 2003; Jacques et al. 2013). Also common are selective breeding and unnatural reproduction. Selective breeding manipulates growth in ways that cause suffering among animals—particularly for chickens, which experience severely decreased mobility as their breasts are genetically engineered by scientists working for the food industry to be too large for their legs to support. Unnatural reproduction is forced upon dairy cows and breeding sows, which are kept perpetually pregnant so that they may continue to produce milk and piglets, respectively. For dairy cows, constantly having to produce milk often results in mastitis—a painful inflammation of the udders. For breeding sows, unnatural, forced breeding means confinement to gestation crates (ASPCA 2013; Jacques et al. 2013). Furthermore, animals are subjected to unclean air caused by a buildup of animal waste in enclosed spaces, lack of veterinary care, and use of unnatural lighting to manipulate growth and behavior (ASPCA 2013).

At the slaughter stage, these animals commonly experience inhumane and ineffective killing (Lawrence 2004). For example, the practices used to render the animal unconscious (e.g., captive bolt gun, electrocution) prior to slaughter are sometimes ineffective and the animal is nonetheless sent on to the next stages of processing while still alive (PETA n.d.). Chickens are hung upside down from shackles and sent assembly-line style past workers or mechanical knives, which slice the chickens' throats; the line moves so rapidly that workers often miss the throat and the animal is mutilated and sent on to the next stages (Kindy 2013; PETA n.d.). For pigs and chickens, later stages of processing include being scalded after slaughter in order to remove hair and feathers. Several undercover videos from pig- and chicken-processing plants have shown that it is not uncommon for pigs and chickens to enter the scald tanks while still alive due to ineffective killing methods (PETA n.d.; Warrick 2001). Indeed, the USDA estimates that approximately 825,000 chickens per year die in the scald tanks (Kindy 2013).

Harm to the Environment and Human Health

Industrial agriculture is not just harmful to animals, but to the environment as a whole and, as a consequence, to humans. These intensive operations consume large amounts of fossil fuels and water, while generating significant water pollution from waste runoff. Intensive farming, particularly intensive animal farming, threatens human health in other ways as well. Illnesses such as *E. coli* and *salmonella* infections have originated with animals raised on industrial farms[2] (Silbergeld et al. 2008), and animal waste has caused outbreaks of cryptosporidiosis in Scotland and England (Centner 2004). In addition, large concentrations of animal waste not only contain pathogens that can lead to illness, but also release dangerously high levels of nitrogen and phosphorus into the air, which can be quite harmful to infants in particular (Centner 2004). Respiratory problems are not uncommon among individuals who work in or live near confined pig operations (Centner 2004; Consumers Union 2000).

We must also consider the fact that animals raised on industrial farms are treated heavily with antibiotics to promote growth and to reduce illness in crowded conditions—approximately 26.6 million pounds of antibiotics are given to farmed animals every year, which is more than eight times more than the amount consumed by humans (Brody 2001). Only about 8 percent of this massive amount of antibiotics is actually used to fight infections (Centner 2004). The extensive prophylactic administration of antibiotics to intensively farmed animals has been found to lead to mutations in bacterial strains as well as a reduction in the efficacy of the drugs for humans who have consumed antibiotic-laden meats (Landers et al. 2012; Silbergeld et al. 2008). Because of these hazards, in January 2006, the European Union (EU) began prohibiting the use of antibiotics in livestock for purposes of enhancing growth and feed efficiency (European Union 2005). In the United States, organizations such as the American Medical Association and the Department of Health and Human Services have stated their opposition to the prophylactic antibiotic use in livestock, but until December 2013, the Food and Drug Administration (FDA) allowed the use of antibiotics as growth enhancers (Mathews 2001; Tavernise 2013). The FDA's latest policy bans the use of antibiotics for the purpose of increasing growth and stipulates that farmers must first procure a prescription from a veterinarian in order to use the drugs to treat sick animals (Tavernise 2013). The changeover to this new policy is expected to be complete in 2017.

In this chapter we have examined many of the general ways that intensive animal farming can be harmful. The next chapter explores specific cases of harm committed against the environment and animals by Tyson Foods.

NOTES

1. Even those industrial farms that are not involved with animal production can be seen to cause harm to the environment. For example, the increase of monocultures—the cultivation of a single crop over consecutive years—leads to soil depletion and unbalanced ecosystems. Monocultures are also more susceptible to pest invasion, which is likely to lead to an overuse of pesticides that are harmful to both the environment and human health (Altieri 2000).

2. Note that one study recently compared 50 conventionally raised beef samples with 50 "grass-fed" beef samples and found no significant differences in contamination rates (Zhang et al. 2010).

REFERENCES

Altieri, Miguel A. 2000. Ecological Impacts of Industrial Agriculture and the Possibilities for Truly Sustainable Farming. In *Hungry for Profit: The Agribusiness Threat to Farmers, Food, and the Environment*, ed. F. Magdoff, J.B. Foster, and F.H. Buttel. New York: Monthly Review.

Animal Legal and Historical Center. 2014. *United States Code Annotated. Title 7. Agriculture. Chapter 48. Humane Methods of Livestock Slaughter.* East Lansing: Animal Legal and Historical Center. http://www.animallaw.info/statutes/stusfd7usca1901.htm. Accessed 15 May 2014.

Animal Legal Defense Fund. 2014. *Farmed Animals and the Law.* San Francisco: Animal Legal Defense Fund. http://aldf.org/resources/advocating-for-animals/farmed-animals-and-the-law/. Accessed 25 Feb 2014.

Animal Welfare Act. 1966. 7 U.S.C. 2131.

ASPCA. 2013. *What Is a Factory Farm?* New York: American Society for the Prevention of Cruelty to Animals. http://www.aspca.org/Fight-AnimalCruelty/farm-animal-cruelty/what-is-afactory-farm. Accessed 18 Aug 2013.

Brody, Jane E. 2001. Studies Find Resistant Bacteria in Meats. *New York Times.* http://www.nytimes.com/2001/10/18/us/studies-find-resistant-bacteria-in-meats.html. Accessed 25 Feb 2014.

Centner, Terence J. 2004. *Empty Pastures: Confined Animals and the Transformation of the Rural Landscape.* Champaign: University of Illinois Press.

Consumers Union. 2000. *Animal Factories: Pollution and Health Threats to Rural Texas.* Yonkers: Consumers Union. http://consumersunion.org/pdf/CAFOforweb.pdf. Accessed 25 Feb 2014.

Delgado, Christopher L. 2003. Rising Consumption of Meat and Milk in Developing Countries Has Created a New Food Revolution. *The Journal of Nutrition* 11: 39075–33105.

European Union. 2005. *Ban on Antibiotics as Growth Promotors in Animal Feed Enters into Effect.* http://europa.eu/rapid/press-release_IP-05-1687_en.htm. Accessed 25 May 2014.

Florer, John H. 1968. Major Issues in the Congressional Debate of the Morrill Act of 1862. *History of Education Quarterly* 8: 457–478.

Friedrich, Bruce. 2013. The Cruelest of All Factory Farm Products: Eggs from Caged Hens. *The Huffington Post.* http://www.huffingtonpost.com/bruce-friedrich/eggs-from-caged-hens_b_2458525.html. Accessed 30 Oct 2016.

Harrison, Ruth. 1966. *Animal Machines: An Expose of "Factory Farming" and Its Danger to the Public.* New York: Ballantine Books.

Hirsch, Veronica. 2003. *Detailed Discussions of Legal Protections of the Domestic Chicken in the United States and Europe.* Animal Legal and Historical Center. https://www.animallaw.info/article/detailed-discussion-legal-protections-domestic-chicken-united-states-and-europe#id-11. Accessed 30 Oct 2016.

HSUS. 2012. *Barren, Cramped Cages.* Washington, DC: Humane Society of the United States. http://www.humanesociety.org/issues/confinement_farm/facts/battery_cages.html#.Uwzys_ldWSo. Accessed 25 Feb 2014.

———. 2014. *Crammed into Gestation Crates.* Washington, DC: Humane Society of the United States. http://www.humanesociety.org/issues/confinement_farm/facts/gestation_crates.html. Accessed 23 May 2014.

————. 2016. *Farm Animal Statistics: Slaughter Totals.* Washington, DC: Humane Society of the United States. http://www.humanesociety.org/news/resources/research/stats_slaughter_totals.html. Accessed 1 Dec 2016.

Humane Methods of Slaughter Act. 1958. 7 U.S.C. 1901.

Jacques, Michelle L., Carole Gibbs, and Louie Rivers. 2013. Confined Animal Feeding Operations. *CRIMSOC: The Journal of Social Criminology* 1: 10–63.

Kindy, Kimberly. 2013. USDA Plan to Speed Up Poultry Processing Lines Could Increase Risk of Bird Abuse. *Washington Post.* https://www.washingtonpost.com/politics/usda-plan-to-speed-up-poultry-processing-lines-could-increase-risk-of-bird-abuse/2013/10/29/aeeffe1e-3b2e-11e3-b6a9-da62c264f40e_story.html?utm_term=.ce60a7e38b21. Accessed 10 Dec 2016.

Land Grant College Act. 1862. 7 U.S.C. 301 et seq.

Landers, Timothy, Bevin Cohen, Thomas E. Wittum, and Elaine L. Larson. 2012. A Review of Antibiotic Use in Food Animals: Perspective, Policy, and Potential. *Public Health Reports* 127: 4–22.

Lawrence, Felicity. 2004. *Not on the Label: What Really Goes into the Food on Your Plate.* London: Penguin.

Mathews, Kenneth H. 2001. *Antimicrobial Resistance and Veterinary Costs in U.S. Livestock Production.* Washington, DC: U.S. Department of Agriculture. http://www.ers.usda.gov/publications/aib-agricultural-information-bulletin/aib766.aspx#.U35eP_ldWSo. Accessed 22 May 2014.

Nurse, Angus. 2013. *Animal Harm: Perspectives on Why People Harm and Kill Animals.* Burlington: Ashgate.

PETA. n.d. *Tortured by Tyson.* Norfolk: People for the Ethical Treatment of Animals. http://www.peta.org/videos/tortured-by-tyson/. Accessed 21 May 2014.

Pew Commission on Industrial Farm Animal Production. 2008. *Putting Meat on the Table: Industrial Farm Animal Production in America.* Philadelphia: The Pew Charitable Trust.

Shapiro, Paul. 2013. Putting Pigs Before Politics. *Huffington Post.* http://www.huffingtonpost.com/paul-shapiro/new-jersey-gestation-crates_b_4145284.html. Accessed 30 May 2014.

Silbergeld, Ellen K., Jay Graham, and Lance B. Price. 2008. Industrial Food Production, Antimicrobial Resistance, and Human Health. *Annual Review of Public Health* 29: 151–169.

Tavernise, Sabrina. 2013. F.D.A. Restricts Antibiotics Use for Livestock. *New York Times.* http://www.nytimes.com/2013/12/12/health/fda-to-phase-out-use-of-some-antibiotics-in-animals-raised-for-meat.html?pagewanted=all. Accessed 25 May 2014.

United States Department of Agriculture. 2012. *Agricultural Statistics 2012.* Washington, DC: United States Government Printing Office. http://www.nass.

usda.gov/Publications/Ag_Statistics/2012/2012_Final.pdf. Accessed 25 Feb 2014.

———. 2014a. *An Act to Establish a Department of Agriculture.* http://www.nal.usda.gov/act-establish-department-agriculture. Accessed 16 Feb 2014.

———. 2014b. *Humane Methods of Slaughter Act.* https://awic.nal.usda.gov/government-and-professional-resources/federal-laws/humane-methods-slaughter-act. Accessed 25 Feb 2014.

Warrick, Joby. 2001. They Die Piece by Piece. *Washington Post.* https://www.uta.edu/philosophy/faculty/burgess-jackson/Warrick,%20They%20Die%20Piece%20by%20Piece%20(2001).pdf. Retrieved 25 Feb 2014.

Zhang, Jiayi, Samantha K. Wall, Xu Li, and Paul D. Ebner. 2010. Contamination Rates and Antimicrobial Resistance in Bacteria Isolated from 'Grass-Fed' Labeled Beef Products. *Foodborne Pathogens and Disease* 7: 1331–1336.

CHAPTER 3

The Nature of Tyson's Harms

Abstract This chapter explores and discusses some noteworthy cases of harm perpetrated by Tyson. Cases include multiple incidents of animal harm recorded via undercover video at Tyson processing plants and contracted farms as well as multiple documented environmental harms that resulted in legal action against Tyson.

Keywords Tyson Foods • Environmental violations • Undercover video

The foregoing discussion of harm caused by industrialized agriculture provides an umbrella under which we may understand the particular harms attributable to Tyson Foods. Any one corporation is only a piece of the puzzle of harm. But Tyson is the largest corporation involved in livestock production and processing (Gazdziak 2013). In this chapter, I will provide an overview of several selected cases of harm perpetrated by Tyson Foods. I examine instances involving harm to the environment and then discuss cases of animal harm.

HARM TO THE ENVIRONMENT

Tyson Foods is responsible for a myriad of environmental harms. In fact, it has been reported that Tyson was the second biggest polluter of US waterways for the years of 2010–2014 (Environment America 2016). In this section, I provide an overview of selected cases of environmental harm perpetrated by Tyson beginning in 2003.

In June 2003, Tyson agreed to pay a total of $7.5 million to the federal government and to the state of Missouri for violating the Clean Water Act—$5.5 million to the federal government, $1 million to the Missouri state government, and $1 million to the Missouri Natural Resources Protection Fund (US Department of Justice 2003). Tyson's chicken-processing plant in Sedalia, Missouri, which processed approximately 1 million chickens per week with hundreds of thousands of gallons of wastewater being produced every day, was first found to be discharging undertreated and untreated wastewater into the Lamine River by an investigator with the Missouri Department of Natural Resources (US Department of Justice 2003). Beginning in 1996, the Missouri Department of Natural Resources cited Tyson several times for their behavior and the state of Missouri filed two lawsuits against Tyson, yet Tyson continued to illegally discharge wastewater (US Department of Justice 2003). Subsequently, the EPA's Criminal Enforcement Division and the FBI conducted an investigation, executing search warrants at the plant in 1999 (Stafford 2003). Tyson stated in its official response that the violations were the result of innocent mistakes by a few employees; however, internal documents revealed that high-level managers were fully aware of the violations (Stafford 2003). In the end, Tyson pleaded guilty to 20 felony violations, admitting to illegally discharging the untreated wastewater (US Department of Justice 2003). The Honorable Howard F. Sachs, the United States District Judge for the United States District Court for the Western District of Missouri, who presided over the case, stated in his order that the fine was small, given Tyson's size and the fact that they blatantly disregarded several warnings and citations. Judge Sachs suggested that Congress increase fines for such cases (Stafford 2003).

In January 2005, Tyson paid $500,000 to settle an air pollution suit brought against the company by the Sierra Club and three Western Kentucky residents (*The New York Times* 2005). The lawsuit was brought on the grounds that four Tyson chicken production facilities in three counties in Western Kentucky were so large that the fumes and dust that came from the farms should be subject to regulation under the Clean Air Act (Bruggers 2002). The complainants argued that the releases from the farm should have been reported but were not, and they alleged that the operations constituted nuisances under state law. In addition to the monetary compensation, Tyson agreed to monitor the air for ammonia and to plant trees as buffers at other chicken farm locations (*The New York Times* 2005).

In April 2013, Tyson paid $3.95 million in fines for violating the Clean Air Act (Jamieson 2013)—eight separate incidents where anhydrous ammonia was accidentally released from refrigeration systems in Tyson plants between 2006 and 2010 (Jamieson 2013). Anhydrous ammonia is an extremely hazardous substance; exposure to it can result in chemical-type burns as well as frostbite since its boiling point is −28 °F (EPA 2006). In addition, anhydrous ammonia is extremely flammable and explosive; "it can be ignited by something as common as the electric flash from a switch" (EPA 2006: 1–2). The first incident occurred in October 2006 at a Tyson plant in South Hutchinson, Kansas—one worker was killed and another was injured by the chemical release (*Joplin Globe* 2013). In November 2006, a release at Tyson's plant in Sedalia, Missouri, caused injuries to three workers (*Joplin Globe* 2013). In December 2006, two incidents occurred—another release at the Hutchison plant caused ten injuries and prompted a full-plant evacuation, and an accidental release occurred at Tyson's Omaha, Nebraska, plant where 5 workers were injured and 475 were evacuated (*Joplin Globe* 2013). Two accidental releases occurred at Tyson's plant in Perry, Iowa— one in November 2007 and another in November 2009 (*Joplin Globe* 2013). One worker was injured in both of these incidents at the Perry plant, with the latter resulting in a 45-day hospitalization due to severe chemical burns and frostbite sustained over 25 percent of his body (Jamieson 2013; *Joplin Globe* 2013). The other accidental releases include incidents at Tyson's Sioux City, Iowa, plant (October 2007), Emporia, Kansas, plant (April 2009), and Cherokee, Iowa, plant (June 2010); one person was injured in each of these incidents (*Joplin Globe* 2013). Inspectors found multiple violations at other Tyson plants in the states where the eight accidents occurred. Overall, 23 Tyson facilities were named in the suit (*Joplin Globe* 2013). In addition to the monetary fine, Tyson agreed to conduct tests of their pipes and to have third-party inspections of their ammonia refrigeration systems in the 23 plants named in the suit (*Joplin Globe* 2013).

The previous three cases were examples of some of the largest fines paid by Tyson for environmental violations, but Tyson has incurred many smaller fines and continues to do so on a regular basis. According to Tyson's 2012 sustainability report,[1] the company incurred $89,701 in fines for the 2010 fiscal year, $154,016 in the 2011 fiscal year, and $18,631 in the 2012 fiscal year (Tyson Foods 2012). Annual fines for Tyson have increased as reported in the 2015 sustainability report. Tyson incurred $3,952,908 in penalties for the 2013 fiscal year, $354, 207 in the 2014 fiscal year, and $403,809 in the

2015 fiscal year (Tyson Foods 2015). In addition, Tyson has revealed in their sustainability reports that they are responsible for numerous reportable chemical releases each year (average of approximately 19 reportable chemical releases per year for fiscal years 2010–2015) (Tyson Foods 2012, 2015). Table 3.1 provides specific information regarding Tyson's environmental fines greater than $5000 for fiscal years 2011–2015.

HARM TO ANIMALS

Tyson Foods has a long record of animal abuse occurring on their suppliers' farms and in their processing facilities. The following overview focuses on cases occurring between 2005 and 2016.

In March 2005, an undercover video recorded by a member of the animal activist group People for the Ethical Treatment of Animals (PETA) emerged depicting widespread abuse in a Tyson chicken-processing plant in Heflin, Alabama (*USA Today* 2005). The video was recorded over a three-month period (December 2004 through February 2005) when the undercover activist was employed by Tyson as a supervisor who was responsible to monitor and prevent chickens from entering scald tanks while still alive (Animal Agriculture Alliance 2014). The video depicted workers killing chickens with their bare hands and chickens being scalded while still alive (PETA n.d.). Tyson responded to the video by stating that the undercover activist was in violation of their animal welfare policies as he allowed the abuse he was hired to prevent. Tyson further claimed that the video had been edited to make it appear as if chickens were being decapitated manually and that these birds had in fact already been cut (*USA Today* 2005). The undercover activist, however, countered that manually decapitating birds was a standard practice in the processing plant and that he had been taught this technique by a supervisor (*USA Today* 2005).

In 2007, PETA conducted an undercover investigation of two additional chicken-processing plants—one in Cumming, Georgia, and one in Union City, Tennessee (Berlin 2008). The investigation resulted in PETA releasing a video that depicted workers urinating on the slaughter line and throwing birds violently into leg shackles. In addition, the video showed one worker admitting to breaking chickens' backs and taking his anger out on the chickens (Animal Agriculture Alliance 2014; Berlin 2008).

In April 2008, an undercover video was recorded by an activist with the Humane Society of the United States (HSUS) at the farm of one of Tyson's hog suppliers, Wyoming Premium Farms in Wheatland, Wyoming (Animal

Table 3.1 Tyson environmental penalties greater than $5000 for fiscal years 2011–2015

Fiscal year	Violation	Location	Penalty
2011	Special order by consent for odor complaints[a]	Harmony, NC	$20,000
2011	Removal of hazardous substances	Mercury Refining Superfund Site, NY	$32,685
2011	Air permit violations	Tenaha, TX	$6656
2011	Failure to immediately report an ammonia release	Vicksburg, MS	$28,800
2011	Failure to immediately report an ammonia release	Logansport, IN	$63,375
2012	Failure to immediately report an ammonia release	Glen Allen, VA	$10,631
2012	Removal of hazardous substances	Marina Shale Processors Superfund Site, LA	$8000
2013	Clean Air Act violations related to refrigeration systems	Facilities in MO, NE, IA, and KS	$3,950,000
2014	Clean Water Act violation—chlorinated water discharge	Sedalia, MO	$5368
2014	Clean Air Act violation (RMP violation)[b]	Albertville, AL	$20,973
2014	Clean Water Act permit violation	New Holland, PA	$5516
2014	Clean Water Act violation—release from sludge storage tank	Harmony, NC	$305,000
2014	Clean Water Act violation—spills from wastewater irrigation system	Madison, NE	$16,000
2015	Clean Water Act violation—feed additive–affected wastewater treatment plant	Monett, MO	$320,000 (plus $220,000 allocated for SEP)[c]
2015	Clean Air Act permit violation	Traverse City, MI	$67,982
2015	Clean Air Act violation (RMP violation)	New Holland, PA	$9042

Data from Tyson Foods' (2012, 2015) sustainability reports
Notes: [a]A special order by consent is a special case where a company applies for an exception to their NPDES permit (North Carolina Department of Environment and Natural Resources n.d.)
[b]Under the Clean Air Act, facilities which hold over a certain amount of a regulated substance must have in place a Risk Management Plan (RMP). RMPs must be updated every five years with the EPA and must include detailed information on accident prevention and emergency response protocols (EPA 2016)
[c]Supplemental Environmental Projects (SEPs) are voluntary environmentally beneficial programs meant to remedy problems related to specific violations. When SEPs are proposed, penalties are often reduced (EPA 2014)

Agriculture Alliance 2014). The undercover activist worked at the farm for approximately one month documenting widespread abuse, including sows being punched and kicked, a worker sitting on a pig with a broken leg, piglets being slammed into the ground, piglets falling through the floor grates into the waste (urine and feces) pits where they were injured or killed, sick piglets being held up in the air and swung around by their hind legs, and numerous pigs with untreated illnesses and wounds (HSUS 2012). Initially, Tyson denied any connection with Wyoming Premium Farms but later admitted that a Tyson-owned subsidiary did purchase hogs from the farm (Zelman 2012a). Tyson stated that they would not be purchasing from Wyoming Premium Farms until they had a chance to investigate (Zelman 2012a). Nine workers from Wyoming Premium Farms were subsequently charged with animal cruelty (Orr 2012; Zelman 2012b). Six of the employees plead guilty to the charges (Biondolillo 2013).

In January 2013, the Animal Legal Defense Fund (ALDF) filed a complaint with the Federal Trade Commission (FTC) against Tyson (Lutz 2013). ALDF's complaint cited false and misleading advertising by Tyson—specifically, statements that appeared on Tyson's website claiming that Tyson Foods was leading the industry in animal welfare (Lutz 2013). Tyson subsequently removed this language from their website and the FTC discontinued their investigation of the complaint based on the changes (Engle 2014).

In November 2013, Mercy for Animals, a US-based nonprofit animal activist group, released a video that documented widespread abuse of pigs on the farm of contracted Tyson hog supplier, West Coast Farms in Oklahoma. An undercover activist with Mercy for Animals recorded the video while he was employed as a farmhand for about a month in 2013 (mid-September through mid-October) (Quirk 2013). This video showed workers kicking, hitting, and throwing pigs, as well as sticking their fingers in pigs' eyes, and, on one occasion, workers throwing a bowling ball at a pig's head (Mercy for Animals n.d.; Quirk 2013). The video also revealed that sick or injured piglets were killed by workers by slamming them on the ground (Quirk 2013). Although blunt force is an industry standard method for euthanizing sick or injured animals,[2] it is recommended that workers ensure that the animal's death is "achieved quickly" (American Association of Swine Veterinarians 2009: 2). The undercover activist observed, however, that piglets were often still alive for hours after they were thrown forcefully onto the ground (Schecter et al. 2013). The undercover activist said that he reported the abuse to the owner of the farm but that the abuse continued and none of the workers indicated that the owner had talked to them about abuse when the activist questioned them about it (Quirk 2013). The owner stated that he was never informed of any abuse and that he terminated all of the employees shown in the video (Quirk 2013). Tyson terminated their contract with West Coast Farms after the video came out, and the Oklahoma Pork Council and the National Pork Producers Council condemned the behaviors depicted in the video (Oklahoma Farm Report 2013). The actions of the workers in the video were also condemned by an expert panel convened by the Center for Food Integrity. The panel, however, defended some of the actions, stating that although they were unpleasant to view, the behaviors were consistent with industry standards (Schecter 2013). Although the District Attorney's Office of Okfuskee County—where West Coast Farms is located—conducted a formal investigation, no formal charges were filed against any of the workers or the farm's owner (Oklahoma Farm Report 2013). In January 2014, Tyson issued a letter to all of their pork suppliers; in this letter, Tyson stated that they would be increasing on-farm audits via their *FarmCheck*™ program[3] (Tyson Foods 2014). The letter also suggested, but did not require, that Tyson pork suppliers install video monitoring in their facilities, that methods other than blunt force be used for euthanizing sick or injured animals, that pain mitigation be used during tail docking and castration procedures, and that housing for pregnant sows be improved (Tyson Foods 2014). The fact that these were suggestions, not requirements, gives the impression of lip service. In terms of housing for pregnant sows, not requiring a phasing out of gestation crates leaves Tyson trailing behind competitors such as Smithfield whose phaseout of gestation crates should be complete in 2017 (HSUS 2014).

Mercy for Animals documented, via undercover video, abuses taking place at Tyson chicken farms and processing plants in multiple states (Mississippi, Tennessee, and Delaware) during 2015 and 2016. They released a video in May 2016 that compiled all of the footage from the various investigations. Given that the abuses were recorded over time and at various locations, Mercy for Animals argued that the abuse cannot be shrugged off as a one-off situation but rather that these instances were indicative of systemic abuse (Mercy for Animals n.d., 2016).

In September 2015, ALDF released undercover footage from a Tyson chicken-processing plant in Texas, which also included an interview with the undercover investigator. Along with the video, ALDF issued a press release which documented complaints they filed with the USDA's Food Safety and

Inspection Service (FSIS), Occupational Safety and Health Administration (OSHA), the Securities and Exchange Commission (SEC), as well as the attorney general of Delaware, the state in which Tyson is incorporated (ALDF 2015).

The animal activist group Compassion over Killing released an undercover video of abuse of chickens at Tyson breeding facilities in Virginia in August 2016. The footage in this video includes workers stomping, kicking, and suffocating birds, and one worker can be heard saying "You can't let nobody see you doing this." In response to the release of the video, Tyson fired the ten workers who appeared in the video and issued a statement that they would retrain all of their bird handlers, though it is unclear if this ever happened (Dicker 2016; Hanson 2016).

Legislation has recently been introduced in several states that would ban undercover audio, video, and photographic documentation on industrial farms (Galli and Kreider 2013). The anti-whistleblower laws that would create such bans are collectively known as ag-gag bills and are lobbied for by agribusiness. Ag-gag bills are one example of how large agribusinesses use their power to perpetuate scenarios that are profitable for them but harmful to animals, humans, and the environment. The next chapter will discuss ag-gag bills in greater depth and will examine corporate power as an explanation for the continuation of the mass harm caused by industrial agriculture. In particular, I discuss my concern in Chap. 4 regarding the construction of the good corporate identity and corporate social responsibility (CSR) and the role of the corporate website as a platform for these constructions.

NOTES

1. A sustainability report is a document published by organizations that assesses the environmental and social impacts of the activities of the organization (Global Reporting Initiative n.d.).
2. In 2013, the American Veterinary Medical Association (AVMA) recommended that those pork producers still using blunt force trauma as their primary method of euthanasia adopt alternative methods (Pork Network 2013).
3. The Tyson *FarmCheck*™ program is an animal well-being program that "includes on-site audits of livestock and poultry farms" (Tyson Foods 2013: 4).

REFERENCES

American Association of Swine Veterinarians. 2009. *On-Farm Euthanasia of Swine: Recommendations for the Producer*. Perry: American Association of Swine Veterinarians.

Animal Agriculture Alliance. 2014. *Truth, Lies and Videotape*. Arlington: Animal Agriculture Alliance. http://www.animalagalliance.org/images/upload/Truth,%20Lies%20and%20Videotape.pdf. Accessed 21 May 2014.

Animal Legal Defense Fund. 2015. *Undercover Investigation Documents Tyson's Cruel, Illegal Treatment of Chickens*. San Francisco: Animal Legal Defense Fund. http://aldf.org/press-room/press-releases/undercover-investigation-documents-tysons-cruel-illegal-treatment-of-chickens/. Accessed 1 Dec 2016.

Berlin, Joey. 2008. PETA on Tyson. *The Emporia Gazette*. http://www.emporiagazette.com/business/article_ae2bc3a5-f888-541d-8ea6-a8d473cbb7ab.html. Accessed 21 May 2014.

Biondolillo, Chelsea. 2013. *Wyoming Premium Farms Employee Found Guilty of Animal Cruelty*. Wyoming Public Media. http://wyomingpublicmedia.org/post/wyoming-premium-farms-employee-found-guilty-animal-cruelty. Accessed 22 May 2014.

Bruggers, James. 2002. Sierra Club Vows Suit over Chicken Farms and the Dust They Produce. *The Louisville Courier-Journal*. http://www.mindfully.org/Farm/Chicken-Farms-Dust-Suit5feb01.htm. Accessed 28 May 2014.

Dicker, Rachel. 2016. Animal Rights Group Records Tyson Foods Employees Horrifically Abusing Chickens. *U.S. News*. http://www.usnews.com/news/articles/2016-08-11/animal-rights-group-records-tyson-foods-employees-horrifically-abusing-chickens. Accessed 1 Dec 2016.

Engle, Mary K. 2014. *Concerning Tyson Foods, Inc's Promotion of the Farm Check™ Animal Well-Being Program*. Federal Trade Commission. http://www.ftc.gov/public-statements/2014/01/letter-mary-k-engle-associate-director-division-advertising-practices. Accessed 21 May 2014.

Environment America. 2016. *America's Next Top Polluter*. Washington, DC: Environment America Research and Policy Center. http://www.environmentamericacenter.org/sites/environment/files/reports/Env_Am_Tyson_v4.pdf. Accessed 12 Apr 2016.

EPA. 2006. *Accident Prevention and Response Manual for Anhydrous Ammonia Refrigeration System Operators*. Washington, DC: Environmental Protection Agency. http://www.epa.gov/region7/toxics/pdf/accident_prevention_ammonia_refrigeration.pdf. Accessed 30 May 2014.

EPA. 2014. *Supplemental Environmental Projects (SEPs)*. Washington, DC: Environmental Protection Agency. Retrieved May 27, 2014, http://www2.epa.gov/enforcement/supplemental-environmental-projects-seps

EPA. 2016. *Risk Management Plan (RMP) Rule Overview*. Washington, DC: Environmental Protection Agency. https://www.epa.gov/rmp/risk-management-plan-rmp-rule-overview. Accessed 22 Dec 2016.

Galli, Cindy, and Randy Kreider. 2013. Ag-gag: More States Move to Ban Hidden Cameras on Farms. *ABC News*. http://abcnews.go.com/Blotter/states-move-ban-hidden-cameras-farms/story?id=18738108. Accessed 18 Aug 2013.

Gazdziak, Sam. 2013. *The National Provisioner's Top 100*. The National Provisioner. http://www.provisioneronline.com/ext/resources/2013May/024-037-top-100.pdf. Accessed 25 Sept 2013.

Global Reporting Initiative. n.d. *About Sustainability Reporting*. https://www.globalreporting.org/information/sustainability-reporting/Pages/default.aspx. Accessed 23 July 2014.

Hanson, Hilary. 2016. Undercover Footage Reveals 'Culture of Animal Cruelty' on Tyson Farms. *Huffington Post*. http://www.huffingtonpost.com/entry/tyson-animal-abuse-chicken-breeding-farm_us_57acbddde4b007c36e4d90f1. Accessed 1 Dec 2016.

HSUS. 2012. *Undercover at a Tyson Supplier: A Humane Society of the United States Investigation*. Washington, DC: Humane Society of the United States. http://www.humanesociety.org/assets/pdfs/farm/Undercover_pigs_050812.pdf. Accessed 21 May 2014.

———. 2014. *Crammed into Gestation Crates*. Washington, DC: Humane Society of the United States. http://www.humanesociety.org/issues/confinement_farm/facts/gestation_crates.html. Accessed 23 May 2014.

Jamieson, Dave. 2013. Tyson Foods Pays $4 Million to Settle Complaint over Worker Exposure to Ammonia. *Huffington Post*. http://www.huffingtonpost.com/2013/04/05/tyson-foods-ammonia_n_3021843.html. Accessed 30 May 2014.

Joplin Globe. 2013. Tyson Agrees to Pay $3.95 Million to Settle Clean Air Act Violations. *Joplin Globe*. http://www.joplinglobe.com/topstories/x237738247/Tyson-agrees-to-pay-3-95-million-to-settle-Clean-Air-Act-violations. Accessed 30 May 2014.

Lutz, Daniel. 2013. *Tyson Exposed by Former Suppliers' Convictions*. San Francisco: Animal Legal Defense Fund. http://aldf.org/blog/tyson-exposed-by-former-suppliers-convictions/. Accessed 21 May 2014.

Mercy for Animals. n.d. *Undercover Investigations: Exposing Animal Abuse*. http://www.mercyforanimals.org/investigations.aspx. Accessed 23 May 2014.

———. 2016. Tyson Foods Slammed Again by Nercy for Animals' Hidden Camera Video Exposing Sickening Animal Abuse. *PR Newswire*. http://www.prnewswire.com/news-releases/tyson-foods-slammed-again-by-mercy-for-animals-hidden-camera-video-exposing-sickening-animal-abuse-300274527.html. Accessed 1 Dec 2016.

North Carolina Department of Energy and Natural Resources. n.d. NPDES Enforcement Actions. Retrieved May 27, 2014, http://portal.ncdenr.org/web /wq/swp/ps/npdes/comp-enf#SOC

Oklahoma Farm Report. 2013. *Oklahoma Hog Farm the Focus of Video of Abusive Treatment of Sows and Pigs.* http://www.oklahomafarmreport.com/wire/news/ 2013/11/06600_HogVideoAbuse11202013_055333.php#.U39B4PldWSo. Accessed 23 May 2014.

Orr, Becky. 2012. 9 Cited in Abuse at Wheatland Pig Farm. *Wyoming News.* http:// www.wyomingnews.com/articles/2012/12/25/news/01top_12-25-12.txt#. U31Fi_ldWSo. Accessed 22 May 2014.

PETA. n.d. *Tortured by Tyson.* Norfolk: People for the Ethical Treatment of Animals. http://www.peta.org/videos/tortured-by-tyson/. Accessed 21 May 2014.

Pork Network. 2013. *AVMA Recommends Alternative to Blunt Force Trauma.* http:// www.porknetwork.com/pork-news/AVMA-recommends-alternative-to-blunt-fo rce-trauma-234790191.html. Accessed 23 May 2014.

Quirk, Mary Beth. 2013. Tyson Foods Breaks Up with Pig Farm After Video Shows Alleged Animal Abuse. *Consumerist.* http://consumerist.com/2013/11/21/ tyson-foods-breaks-up-with-pig-farm-after-video-shows-alleged-animal-abuse/. Accessed 23 May 2014.

Schecter, Anna. 2013. Expert Panel Says Undercover Video Shows Abuse at Pig Farm. *NBC News.* http://www.nbcnews.com/news/other/expert-panel-says-undercover-video-shows-abuse-pig-farm-f2D11637736. Accessed 23 May 2014.

Schecter, Anna, Monica Alba, and Lindsay Perez. 2013. Tyson Foods Dumps Pig Farm After NBC Shows Company Video of Alleged Abuse. *NBC News.* http:// www.nbcnews.com/news/other/tyson-foods-dumps-pig-farm-after-nbc-sho ws-company-video-f2D11627571. Accessed 24 May 2014.

Stafford, Margaret. 2003. Tyson Foods Pleads Guilty to Violating Clean Water Act. *Southeast Missourian.* http://www.semissourian.com/story/112750.html. Accessed 27 May 2014.

The New York Times. 2005. Company News; Tyson Foods Settles Air Pollution Suit for $500,000. *The New York Times,* January 28. http://query.nytimes.com/gst/full page.html?res=9E02EFDE143BF93BA15752C0A9639C8B63. Accessed 28 May 2014.

Tyson Foods. 2012. *Tyson Foods 2012 Sustainability Report Executive Summary.* http://www.tysonsustainability.com/~/media/Sustainability/Files/2012% 20Sustainability%20Report%20Executive%20Summary.ashx. Accessed 27 May 2014.

———. 2013. *Fiscal 2013 Fact Book.* http://ir.tyson.com/files/doc_downloads/ Tyson%202013%20Fact%20Book.pdf. Accessed 17 Apr 2014.

————. 2014. *Tyson Foods Letter to Hog Farmers.* http://www.tysonfoods.com/ Media/News-Releases/2014/01/Tyson-Foods-Letter-to-Hog-Farmers.aspx. Accessed 23 May 2014.

————. 2015. *Our 2015 Sustainability Report.* www.tysonsustainability.com. Accessed 17 Dec 2016.

US Department of Justice. 2003. *Tyson Pleads Guilty to 20 Felonies and Agrees to Pay $7.5 Million for Clean Water Act Violations.* Washington, DC: US Department of Justice. http://www.justice.gov/opa/pr/2003/June/03_enrd_383.htm. Accessed 27 May 2014.

USA Today. 2005. *Tyson, PETA Clash over Chicken Slaughter,* May 25. http://usa today30.usatoday.com/money/industries/food/2005-05-25-tyson-peta_x.htm? csp=15. Accessed 21 May 2014.

Zelman, Joanna. 2012a. Wyoming Premium Farms Abuse Alleged by Humane Society. *Huffington Post,* May 8. http://www.huffingtonpost.com/2012/05/08/ wyoming-premium-farms-abuse-humane-society_n_1499707.html. Accessed 21 May 2014.

————. 2012b. Wyoming Premium Farms Employees Charged with Animal Cruelty, Humane Society Says. *Huffington Post,* December 24. http://www.huffingtonpost. com/2012/12/24/wyoming-premium-farms-charge-animal-cruelty_n_2359465. html. Accessed 21 May 2014.

CHAPTER 4

Contextualizing the "Socially Responsible" Corporation and the Cultural Legitimation of Harm

Abstract Corporate harms and their legitimation are situated within a complex cultural, structural, and historical landscape. This chapter is an effort to illuminate that landscape. The unifying argument of this chapter is that Tyson's harm/socially responsible discourse reflects general attitudes about harm to nonhumans and corporate power, as well as weak corporate regulation. In addition, Tyson's harm/discourse cannot be understood without also understanding the history of corporate public relations or "spin," and its contemporary conduit par excellence, the corporate web page, and the particularly modern "need" for companies to project social responsibility.

Keywords Corporate harm • Dominant social paradigm • Public relations • Corporate identity

Corporate harms and their legitimation are situated within a complex cultural, structural, and historical landscape. This chapter is an effort to illuminate that landscape.

The unifying argument of this chapter is that Tyson's harm/socially responsible discourse reflects general attitudes about harm to nonhumans and corporate power, as well as weak corporate regulation. In addition, Tyson's harm/discourse cannot be understood without also understanding the history of corporate public relations (PR) or "spin," and its contemporary conduit par excellence, the corporate web page, and the particularly modern "need" for companies to project social responsibility.

I first discuss the dominant social paradigm and how this paradigm perpetuates corporate power and results in weak corporate regulatory systems. Then, I turn to a discussion of PR to clarify how corporations persuade the public to think certain ways about their products and the corporations themselves. Corporations construct themselves as socially responsible via various media vectors, including television, print, and the Internet. I will conclude with a brief discussion of the corporate website which is increasingly an indispensable tool for communicating messages to the public.

THE DOMINANT SOCIAL PARADIGM AND CORPORATE POWER

The concept of the dominant social paradigm, "the values, metaphysical beliefs, institutions, habits, etc. that collectively provide social lenses through which individuals and groups interpret their social world" (Milbrath 1984: 7), provides some insight into how harmful practices perpetrated by corporations often go unchallenged by society at large. The dominant social paradigm operates as an underlying logic that directs the everyday actions and interactions of citizens. The term was first introduced by Pirages and Ehrlich (1974) to specifically describe the cultural perspective that permits the widespread degradation of the environment. The current dominant social paradigm in the United States continues to be a cultural perspective that promotes harm to animals and the environment.

Cable (2012) outlines six main logics of the current dominant social paradigm. The first holds that the value of nature lies in the extent to which it can be exploited for production. The second logic deals with circles of compassion—any "other" such as other species, other people, and other generations are excluded from circles of compassion and are thus exploitable. Third, the dominant social paradigm places great importance on consumerism—the ability to consume more is equated to having a good life and being a good citizen. Fourth is the idea of acceptable risks and the belief that the market will regulate itself. Fifth is the notion that growth is limitless and that any environmental problems associated with ever-increasing growth can and will be solved by human ingenuity. Finally, the dominant social paradigm contains the idea that current societal arrangements are not problematic; "*society, culture and politics are basically OK*" (Cable 2012: 123, emphasis in the original). The persistence of this paradigm reflects a merging of the logic of the corporation, to achieve ever-increasing profit, and the logic of the public sphere, to consume products at ever-increasing rates (Allen 2005). These complementary logics have respectively been referred to as the treadmill of production (Schnaiberg 1980) and the treadmill of consumption (Bell 1998). Many scholars note that the dominant social paradigm is not only detrimental to the environment, but that, as part of a larger cultural orientation influenced by neoliberal ideology, it also erodes democracy (Allen 2005; Brisman 2013; Cable 2012; Carey 1997). Cable eloquently notes:

> With the public inculcation of corporate ideology, corporate values and democratic values are conflated in an insidious form of social control: social control by stealth. We are sold the ideology like brand-new, prewashed, already torn blue jeans. Corporations downsize democracy and score a "two-fer": management of public attitudes plus legitimation of the corporate state. (Cable 2012: 124)

Indeed, corporate power and the dominant social paradigm have flourished in the context of neoliberal ideology—and all are mutually reinforcing.

The power enjoyed by corporations is legitimized within the dominant social paradigm; critical criminologists Henry and Milovanovic note that "corporations as excessive expressors of power ... will inevitably attempt to constitute law to reflect the legitimacy of their power and the illegitimacy of others" (1996: 117). Indeed, corporate actors have a vested interest in making sure that the practices in which they engage occur within the bounds of the law and so are inclined to influence the law to their favor. Studies of corporate influence on the law have shown that corporations utilize their resources in "concerted attempt[s] to prevent their socially injurious behaviors from being criminalized" (Box 1983). Even in instances where the state has been successful in creating laws that restrict behaviors of corporate actors, enforcement of the laws falls to relatively toothless regulatory agencies. Of the regulatory agencies in the United States and the United Kingdom, Box makes four points that are telling of the limits of their power:

1. Regulatory agencies do have the power to "initiate or recommend criminal prosecution," but "they are primarily designed to be regulatory bodies whose main weapon against corporate behavior is administrative" (1983: 45).

2. The financial resources of the regulatory agencies are paltry when compared to the vast amounts of wealth available to national and multinational corporations.
3. Related to the second point, regulatory agencies do not possess anywhere near equivalent legal resources to national and multinational corporations. That is to say, the government lawyers, whom Box refers to as "all-rounders," who would represent the state in court proceedings, are no match for the specialized corporate lawyers who have spent inordinate amounts of time focusing on very specific areas and finding loopholes (1983: 46).
4. Transnational corporations can easily relocate their base of operations, or "their illegal activities, at least" to other countries where regulations are nonexistent or are more lax (1983: 46).

The observations made by Box over three decades ago still hold true today. Tyson, as a large corporation, enjoys legitimation through the dominant social paradigm and is able to manifest their power to influence the law and access resources that dwarf those of the regulatory agencies that would sanction any violations.

The political economy contexts that enable harmful practices to continue operate alongside the cultural legitimation of the corporation and their power/actions. Indeed, those political economy contexts help to create the cultural legitimation; they are mutually reinforcing. A consenting majority of citizens must allow the practices of the corporation to continue. As Cable notes: "The social institutions that reflect economic imperatives are bound, chicken-and-egg-style, with culture... Culture normalizes and legitimates social institutions" (2012: 122). Certainly, the consent of the public comes in part from the simple fact that we understand what the corporation is doing as operating within the bounds of the law, which is legitimate by definition. But the legality of the actions of the corporation is only part of the story. In order to get people on board with them and to induce people to purchase their products or pay for their services, corporations must turn to persuasive appeals and other marketing strategies. The persuasive appeals used in marketing are not just about selling a product anymore, but are more about selling a lifestyle, an image or an idea (see Brisman and South 2013, 2014). Indeed, corporations sell their brand images, which are designed to fit the desired self-image of target consumers (Dowling 2001). Corporations communicate their identity and their brand not only through traditional advertising campaigns, but also via public relations. Although both advertising/marketing and PR are used to communicate ideas to the public, they are generally distinct practices. Advertising or marketing are instances where a company has *paid* for time, whereas PR often occurs under the guise of press releases and other communication with the public via media time that has not been directly purchased. Beder (2002) notes:

> The art of PR is to 'create news'; to turn what are essentially advertisements into a form that fits news coverage and makes a journalists' job easier while at the same time promoting the interests of the client. (113)

As such, most large companies now spend more on PR than they do on traditional advertising (Beder 2002). An in-depth discussion of PR follows in the next section.

PUBLIC RELATIONS

PR is the organizational practice of managing and disseminating information to the public (Grunig and Hunt 1984). PR has become a multibillion dollar industry with the top 20 firms in the United States bringing in more than $1.5 billion in 2013 (O'Dwyer 2014). Scholars have traced PR back to ancient societies, with Edward Bernays, a pioneer of modern PR in both theory and practice, stating, "The three main elements of public relations are practically as old as society: informing people, persuading people, or integrating people with people" (1952: 12). Modern PR, however, is usually attributed to either Bernays or Ivy Lee, discussed below.

Ivy Lee was one of the founders of the first PR agency, the Publicity Bureau, founded in 1900, and is widely considered to have created the press release (Diggs-Brown 2011). The press release includes such things as "news, feature stories, bulletins and other announcements which flood media offices" (Beder 2002: 112). Lee's first PR client was the Pennsylvania Railway, which had built a reputation of secrecy surrounding accidents as they regularly denied access to reporters when accidents occurred (Harrison and Moloney 2004). Lee approached this lack of public trust by inventing the press release, which, while avoiding alienating reporters, ensured that the company itself managed media coverage of any accidents (Beder 2002; Harrison and Moloney 2004). The use of the press release and Lee's insistence that the Pennsylvania Railway be more forthcoming about accidents resulted in the restoration of the railway's reputation (Harrison and

Moloney 2004). During Lee's time with the railway, an accident took place on the New York Central Railroad just weeks following an accident on the Pennsylvania Railway. While New York Central Railroad maintained the position of denying access to reporters, Lee invited reporters to Pennsylvania Railways' accident site and answered all of their questions. Subsequently, columns and editorials praising Pennsylvania Railway and condemning New York Central began to appear (Harrison and Moloney 2004). In 1906, while Lee was employed by the coal mining company, George F. Baer and Associates, he came under fire as the public aligned with striking workers during a labor dispute. The criticisms led Lee to release a "Declaration of Principles" to all city newspaper editors (Harrison and Moloney 2004). At the heart of these principles were the ideas of transparency and honest communication (Diggs-Brown 2011).

Whereas Lee focused on informing the public, Bernays, a nephew of Sigmund Freud, devised elaborate modes of managing opinions and persuading people by applying psychological principles to PR (Bernays 1923). Bernays both theorized about and practiced PR, and he expanded the PR repertoire to include photo-ops, press conferences, and the like, to create staged events on which the media could report (Beder 2002). Bernays famously served as part of President Woodrow Wilson's propaganda team, which was responsible for the campaign that shaped public opinion on the United States' entry into World War II (Carey 1997). After the war, Bernays was able to transfer the same principles used in the war campaign to business campaigns. Of this shift, Bernays wrote:

> Businessmen, private institutions, great universities—all kinds of groups— became conditioned to the fact that they needed the public; that the public could now perhaps be harnessed to their cause as it had been harnessed during the war to the national cause, and that the same methods could be used. (1952: 78)

Indeed, Bernays was concerned with manipulating the thought processes of consumers. While working in the private sector, one of his more famous campaigns was a PR campaign for Lucky Strike cigarettes. The goal of Bernays' campaign was to create acceptance of women smoking in public as it was thought that this could potentially double the market (Amos and Haglund 2000). Part of the campaign involved hiring women to walk in New York City's Easter Parade smoking "torches of freedom" (Amos and Haglund 2000). By casting cigarettes as symbols of women's independence in this early instance of what might now be called undercover or stealth marketing,[1] some of the social stigma of women smoking in public would be diminished.

In sum, PR is still designed to persuade the public to think in certain ways about companies and their products. The sheer number of outlets with which corporations may disseminate their message to the public, however, has greatly increased since the early days of PR with the expansion of media outlets.

Certainly, the general public does not consist of automatons just waiting to be told what to think. Indeed, the public does have power to influence the behavior of corporations. Corporations must constantly update their images and branding in order to meet the changing views of consumers. For example, there has been a growing interest in buying products that meet certain ethical standards and thus we have seen increases in these types of products, including hybrid and electric vehicles, fair-trade and organic food and clothing. In the next section, I discuss the rise of ethical consumerism and the effect that it has had on corporations' projections of social responsibility.

ETHICAL CONSUMERISM, CORPORATE IDENTITY, AND CSR

Corporations' enormous drive for profit is in direct conflict with the idea of sustainability,[2] which, biologically speaking, refers to an ecosystem's ability to remain active and diverse. In other words, sustainability concerns the survival of every living thing on earth. When our dominant social paradigm fails to problematize environmental destruction, the continued survival of the planet and thus every species on the planet is threatened. However, regardless of whether consumption and sustainability are incompatible, the idea of sustainability has crept into the popular conscience of consumers, due in part to a greater overall awareness of environmental issues (Strong 1996); it is becoming increasingly important to consumers as they attempt to shop for a better world. This concern for sustainability and the general social contributions of companies is known as "ethical consumerism" (Irving et al. 2002). The ethical consumer is concerned with a multiplicity of issues including animal welfare, environmental welfare, and human rights (Irving et al. 2002; Strong 1996). The demands of ethical consumers have brought about some change in corporate practices. For example, Jack Daniels stopped sponsoring angling competitions after a successful boycott campaign and it was a consumer boycott that led to dolphin-safe practices (and labels) in the tuna trade (Irving et al. 2002). But perhaps the biggest impact that ethical consumerism has had on corporations is a more general one concerning images of goodness. In order to maintain customer loyalty in the era of ethical consumerism, corporations must communicate the idea to consumers that they are worthy and must present themselves in such a way that consumers and others (e.g., potential employees, shareholders) would want to align with them and buy their products or services.

Such positive corporate identity work becomes problematic when image is distant from reality—especially when the image distorts or conceals a reality of harm-doing. Companies that harm animals and the environment (including legal harms) are aware that their practices do not sit well with various publics, especially consumers who are vital to the financial success of the corporation (Van Riel and Balmer 1997). Therefore, organizations must figure out how to legitimize their actions to maintain consumer trust and loyalty.

One of the strategies employed by corporations is "greenwashing," which can be understood as the use of advertising and PR campaigns that "affect the environmental consciousness of consumers," misleading them in terms of a company's environmental practices and/or the environmental benefits of specific products (Greenpeace 2013; Lynch and Stretesky 2003: 221). Corporate greenwash distorts the reality of corporate actions while simultaneously suppressing dissenting voices (Holcomb 2008). Consider, for example, Fiji water, which Greenlife Online named as one of the top five worst greenwashers of 2012. Fiji has claimed in advertisements that every drop of their water is "green" and that the company actually has a "negative carbon footprint." This claim is juxtaposed against the reality that their product is quite harmful to the environment as it intensively uses fossil fuels in production and shipping (Gang 2012). Holcomb (2008) examined the Internet as an outlet for corporations to engage in greenwashing, noting that corporate websites present information in an authoritative and professional way that serves to confuse individuals attempting to learn the "truth" about a corporation's practices. The utilization of corporate websites by those engaged in intensive farming to promulgate an image of good corporate citizens who are "stewards of the land, animals and environment" (Tyson Foods 2013) could be considered a form of greenwashing.

Although most people approve of the use of animals for food and medical research (Driscoll 1992), many also want animals used for food to be treated in humane ways that minimize harms (Lusk et al. 2007). And so it follows that the powerful corporations of agribusiness have a vested interest in masking their harmful actions. In order to distract from their harmful actions, corporations must engage in many practices to obscure their involvement to the public. For example, the physical sites of food production, including the farms themselves, slaughterhouses and meat-packing plants, are purposely situated in remote locations to maintain distance from the everyday public (Plous 1993). Further, legislation has recently been introduced in several states that would ban the use of photo and videographic equipment at industrial farms (Galli and Kreider 2013). Kansas and Montana were early adopters of these so-called ag-gag, passing such legislation in 1990 and 1991, respectively. Then in the early to mid-2010s, there seemed to be an explosion of ag-gag bills introduced across the country. Arizona, Arkansas, California, Florida, Idaho, Illinois, Indiana, Iowa, Kentucky, Minnesota, Missouri, Nebraska, New Hampshire, New Mexico, New York, North Carolina, North Dakota, Pennsylvania, Tennessee, Utah, Vermont, Washington, and Wyoming all introduced some form of ag-gag legislation between 2011 and 2015. Many of the bills were defeated (including 11 defeated in 2013 alone), but those in Iowa Missouri, Utah, North Dakota, South Carolina, and Arkansas were passed during this time (ASPCA 2016; Flynn 2013). It is likely that many of the defeated bills will be reintroduced like the Arkansas bill, which was one of the 11 bills defeated in 2013 but then subsequently passed in 2014 after being reintroduced (Animal Legal and Historical Center 2014; Arkansas State Legislature 2014; Flynn 2013).

Corporations involved in food production, like other corporations, must also engage in selling their image to a public who is increasingly demanding more humane treatment of animals that are raised for food. The idea of CSR is not new but corporations are nowadays taking it more seriously than they have in the past. Indeed, until recently, many believed that the only responsibility of the corporation was to accumulate ever-increasing profits, and, legally speaking, the purpose of the corporation continues to be the maximization of profits for shareholders (Hartman et al. 2008). For example, the economist Milton Friedman wrote that corporations exercising social responsibility would in effect force people to contribute to some cause that they may not necessarily want to contribute to: "those who favor the taxes and expenditures in question have failed to persuade a majority of their fellow citizens to be of like mind and...they are seeking to attain by undemocratic procedures what they cannot attain by democratic procedures" (1970: 123). This view of profits as the sole responsibility of the corporation has been changing among certain segments of the public. As increasing numbers of employees, clients/consumers, and shareholders place value on the civic behaviors of the companies they hold stock in/or patronize, expectations of transparency and disclosure of corporate actions have also increased (Capriotti and Moreno 2007; Deegan 2002). Capriotti and Moreno conceptualize corporate responsibility as an organization's commitment to "information transparency and ethical behavior" in all aspects of its operations including product development and production, communications with the public, and in the management of the company (2007: 85). Carroll (1991) presents a model of corporate social responsibility as a pyramid with economic responsibilities as the base (traditional responsibility of the company to earn profits for shareholders), legal obligations as the next level of the pyramid (complying with regulations), ethical responsibilities (to do what is right, just and fair) above that, and philanthropic responsibilities (contributing time and money to society) at the top of the pyramid. The growing importance of corporate social responsibility, coupled with the fact that a large percentage of people believe that businesses do not make enough of an effort to be socially responsible (Hartman et al. 2008), makes it vital for corporations to impress upon the public that they are good corporate citizens.

Hartman et al. (2008) outline three distinct models of corporate social responsibility: the *corporate citizenship model*, the *social contract model*, and the *enlightened self-interest model*. The *corporate citizenship model* is the idea that engagement in corporate social responsibility is done only for the public good. In the corporate citizenship model, the corporation does not expect any returns for their efforts and hold that because they have the ability to do good, they *should* do good (Hartman et al. 2008). Companies that utilize the corporate citizenship model often have close ties to the communities in which they operate.

The *social contract model* holds that it is the responsibility of the corporation to respect "the moral rights of various stakeholders" (Hartman et al. 2008: 150). In this model, social responsibility is seen as an obligation of the corporation: an unspoken agreement exists between the public and businesses where the public allows the company to remain successful so long as they fulfill their obligation of being socially responsible.

Finally, the *enlightened self-interest model* essentially suggests that engaging in corporate responsibility gives a company a competitive edge by incorporating the image of responsibility into the organization's brand. In this model, the rationale for corporate social responsibility is risk reduction, market reputation, brand image, stakeholder relationships, and the long-term interests of the corporation (Hartman et al. 2008).

It may be the case that corporations like Tyson try to convey to the public not only that they are socially responsible but also that they are socially responsible for selfless reasons, aligning with the corporate citizenship model. Regardless of which model a corporation utilizes, however, social responsibility is something that resonates with the public and so this must somehow be communicated to the public. As such, the corporate website has become a vital tool in the information age—a time that is characterized by the unfettered transfer of information and by speed of access to knowledge that was not previously possible. As Castells notes:

> The shift from traditional mass media to a system of horizontal communication networks organized around the Internet and wireless communication has introduced a multiplicity of communication patterns at the source of a fundamental cultural transformation, as virtuality becomes an essential dimension of our reality. (2010: xviii)

In the information age, creating and maintaining a web presence is of vital importance for corporations (Campbell and Beck 2004; Capriotti and Moreno 2007; Heinze and Hu 2006). Indeed, any corporation or business without a website is likely to be considered behind the times. Corporations use their websites for many things, including to sell products online, to provide customer service, and as a platform to communicate their ideas to the public (Capriotti and Moreno 2007). In particular, the corporate website has become an important tool for communicating the corporate identity and corporate social responsibility as interest in ethical consumerism has increased (Capriotti and Moreno 2007; Deegan 2002). Thus, the corporate website becomes a good point of entry for researching how corporations discursively construct their identities and their actions.

Summary

Today's large corporations follow a tradition of public relations, dating at least to the turn of the twentieth century, for the purpose of culturally legitimizing their actions including their harmful actions. They find a receptive audience in Western countries given the dominant social paradigm, which holds that corporations are trustworthy and that status quo depletion and destruction of nonhuman animals and the environment are permissible.

The dominant social paradigm is the cultural basis of complacency in the face of such harm. But the idea of a merely complacent public is at odds with popular preferences for social responsibility and ethical consumerism. Thus, corporations that cause harm to nonhumans and the ecological environment, rather than resting on blind trust from the majority of citizens, must generate messages and stories that project care and moral decency. Increasingly, such messages and stories are disseminated through their websites. Along with Facebook and other social media interfaces, websites are the coin of the realm of public relations in the twenty-first century. Tyson Foods is but one corporation embedded within the broader cultural context discussed in this chapter. But, as the largest producer of meat products based in the United States, the messages produced by Tyson Foods on their website provide insight as to just which discourses permit widespread animal suffering and environmental degradation.

NOTES

1. Undercover or stealth marketing refers to the corporations' strategic hiring of individuals to give positive reviews of products or to use products in public. For example, Blackberry's undercover marketing campaign involved hiring attractive women to ask strangers to take their picture or to ask men to put their phone numbers into the smartphone (Osterhout 2010).
2. Other theoretical frames have been developed for understanding the relationship between consumption and sustainability. For example, the ecological modernization perspective conceptualizes environmental harm as stemming from the process of modernization—industrialization in particular. The perspective, however, holds that it is within the context of modernity that we will be able to solve our environmental problems; human ingenuity will remedy those problems by producing some kind of technological fix (see Buttel 2000).

REFERENCES

Allen, David S. 2005. *Democracy Inc.: The Press and Law in the Corporate Rationalization of the Public Sphere.* Urbana: University of Illinois Press.

Amos, Amanda, and Margaretha Haglund. 2000. From Social Taboo to 'Torch of Freedom': The Marketing of Cigarettes to Women. *Tobacco Control* 9: 3–8.

Animal Legal and Historical Center. 2014. *Code of Laws of South Carolina 1976 Annotated. Title 47. Animals, Livestock and Poultry. Chapter 21. Farm Animal and Research Facilities Protection Act.* East Lansing: Animal Legal and Historical Center. http://www.animallaw.info/statutes/stussc47_21_10.htm. Accessed 19 Apr 2014.

Arkansas State Legislature. 2014. *SB13- Providing Legal Protection to Animal Owners and Their Animals and to Ensure That Only Law Enforcement Agencies Investigate Charges of Animal Cruelty.* http://www.arkleg.state.ar.us/assembly/2013/2013R/Pages/BillInformation.aspx?measureno=SB13. Accessed 19 Apr 2014.

ASPCA. 2016. *Ag-Gag Legislation by State.* New York: American Society for the Prevention of Cruelty to Animals. http://www.aspca.org/animal-protection/public-policy/ag-gag-legislation-state. Accessed 29 Oct 2016.

Beder, Sharon. 2002. *Global Spin: The Corporate Assault on Environmentalism.* White River Junction: Chelsea Green Publishing.

Bell, Michael M. 1998. *An Invitation to Environmental Sociology.* Thousand Oaks: Pine Forge Press.

Bernays, Edward. 1923. *Crystallizing Public Opinion.* New York: Boni and Liveright.

———. 1952. *Public Relations.* Norman: University of Oklahoma Press.

Box, Steven. 1983. *Power, Crime, and Mystification.* London: Tavistock.

Brisman, Avi. 2013. Not a Bedtime Story: Climate Change, Neoliberalism, and the Future of the Arctic. *Michigan State International Law Review* 22: 241–289.

Brisman, Avi, and Nigel South. 2013. A Green-Cultural Criminology: An Exploratory Outline. *Crime Media Culture* 9: 115–135.

———. 2014. *Green Cultural Criminology: Constructions of Environmental Harm, Consumerism, and Resistance to Ecocide.* London/New York: Routledge.

Buttel, Frederick H. 2000. Ecological Modernization as Social Theory. *Geoforum* 31: 57–65.

Cable, Sherry. 2012. *Sustainable Failures: Environmental Policy and Democracy in a Petro Dependent World.* Philadelphia: Temple University Press.

Campbell, David, and A. Cornelia Beck. 2004. Answering Allegations: The Use of the Corporate Website for Restorative Ethical and Social Discourse. *Business Ethics: A European Review* 13: 100–116.

Capriotti, Paul, and Angeles Moreno. 2007. Corporate Citizenship and Public Relations: The Importance and Interactivity of Social Responsibility Issues on Corporate Websites. *Public Relations Review* 33: 84–91.

Carey, Alex. 1997. *Taking the Risk Out of Democracy: Corporate Propaganda Versus Freedom and Liberty.* Champaign: Illini Books.

Carroll, Archie B. 1991. The Pyramid of Corporate Social Responsibility: Toward the Moral Management of Organizational Stakeholders. *Business Horizons* 34: 39–48.

Castells, Manuel. 2010. *The Rise of the Network Society.* Chichester: Wiley Blackwell.

Deegan, Craig. 2002. The Legitimizing Effect of Social and Environmental Discourses- A Theoretical Foundation. *Accounting, Auditing, and Accountability Journal* 15: 282–311.

Diggs-Brown, Barbara. 2011. *Strategic Public Relations: An Audience-Centered Practice.* Independence: Cengage.

Dowling, Grahame. 2001. *Creating Corporate Reputations: Identity, Image and Performance.* New York: Oxford University Press.

Driscoll, Janis W. 1992. Attitudes Toward Animal Use. *Anthrozoös* 5: 32–39.

Flynn, Dan. 2013. 2013 Legislative Season Ends with Ag-Gag Bills Defeated in 11 States. *Food Safety News.* http://www.foodsafetynews.com/2013/07/2013-legislative-season-ends-with-ag-gag-bills-defeated-in-11-states/#.UgmPT220R4M. Accessed 18 Aug 2013.

Friedman, Milton. 1970. The Social Responsibility of Business Is to Increase Its Profits. *New York Times Magazine.* http://graphics8.nytimes.com/packages/pdf/business/miltonfriedman1970.pdf. Accessed 12 Apr 2014.

Galli, Cindy, and Randy Kreider. 2013. Ag-gag: More States Move to Ban Hidden Cameras on Farms. *ABC News.* http://abcnews.go.com/Blotter/states-move-ban-hidden-cameras-farms/story?id=18738108. Accessed 18 Aug 2013.

Gang, Jeff. 2012. *Don't Be Fooled.* Boston: The Green Life Online. http://thegreenlifeblog.files.wordpress.com/2012/04/dbf-report.pdf. Accessed 18 Aug 2013.

Greenpeace. 2013. *Greenwashing.* Washington, DC: Greenpeace. http://www.stopgreenwash.org/. Accessed 18 Aug 2013.

Grunig, James E., and Todd Hunt. 1984. *Managing Public Relations.* New York: Holt, Rinehart, and Winston.

Harrison, Shirley, and Kevin Moloney. 2004. Comparing Two Public Relations Pioneers: American Ivy Lee and British John Elliot. *Public Relations Review* 30: 205–215.

Hartman, Laura P., Joseph DesJardins, and Chris MacDonald. 2008. *Business Ethics: Decision Making for Personal Integrity and Social Responsibility.* 3rd ed. New York: McGraw-Hill.

Heinze, Nathan, and Hu Qing. 2006. The Evolution of Corporate Web Presence: A Longitudinal Study of Large American Companies. *International Journal of Information Management* 26: 313–325.

Henry, Stuart, and Dragan Milovanovic. 1996. *Constitutive Criminology: Beyond Postmodernism.* London: Sage.

Holcomb, Jeanne. 2008. Environmentalism and the Internet: Corporate Greenwashers and Environmental Groups. *Contemporary Justice Review* 11: 203–211.

Irving, Sarah, Rob Harrison, and Mary Raynor. 2002. Ethical Consumerism: Democracy Through the Wallet. *Journal of Research for Consumers* 3: 63–83.

Lusk, Jayson L., F. Bailey Norwood, and Robert W. Prickett. 2007. *Consumer Preferences for Farm Animals Welfare: Results of a Nationwide Telephone Survey*. Washington, DC: American Farm Bureau Federation. http://asp.okstate.edu/baileynorwood/Bailey/Research/InitialReporttoAFB.pdf. Accessed 22 May 2014.

Lynch, Michael J., and Paul B. Stretesky. 2003. The Meaning of Green: Contrasting Criminological Perspectives. *Theoretical Criminology* 7: 217–238.

Milbrath, Lester W. 1984. *Environmentalists: Vanguard for a New Society*. Albany: SUNY Press.

O'Dwyer. 2014. *Worldwide Fees of Top PR Firms with Major U.S. Operations*. http://www.odwyerpr.com/pr_firm_rankings/independents.htm. Accessed 14 Apr 2014.

Osterhout, Jacob E. 2010. Stealth Marketing: When You're Being Pitched and You Don't Even Know It. *New York Daily News*. http://www.nydailynews.com/lifestyle/stealth-marketing-pitched-don-article-1.165278. Accessed 22 June 2014.

Pirages, Dennis, and Paul R. Ehrlich. 1974. *ARK II*. New York: W.H. Freeman and Company.

Plous, Scott. 1993. Psychological Mechanisms in the Human Use of Animals. *Journal of Social Issues* 49 (1): 11–52.

Schnaiberg, Allan. 1980. *The Environment: From Surplus to Scarcity*. New York: Oxford University Press.

Strong, Carolyn. 1996. Features Contributing to the Growth of Ethical Consumerism: A Preliminary Investigation. *Marketing Intelligence and Planning* 14: 5–13.

Tyson Foods. 2013. *Core Values*. http://www.tysonfoods.com/Our-Story/Core-Values.aspx. Accessed 11 Aug 2013.

Van Riel, Cees B.M., and John M.T. Balmer. 1997. Corporate Identity: The Concept, Its Measurement and Management. *European Journal of Marketing* 31: 340–355.

CHAPTER 5

Disguising Harms: Talking and Not Talking About *It*

Abstract This chapter discusses how Tyson disguises their actions toward animals and the environment. Unexamined, the passages from Tyson's website that discuss the environment and animal well-being give the impression that Tyson is proactive and transparent. On closer inspection, we find that Tyson says a lot about their *beliefs*, but very little about what they are actually doing.

Keywords Verb processes • Concealing action • Metonymy

How does Tyson manage to remain the largest player in meat production despite their troubling record of environmental violations and harm to animal welfare? In this chapter, I discuss the findings of my analysis, which addresses that question in terms of Tyson's discursive self-construction. I found that the discourses on Tyson's website accomplish three main tasks: they *disguise* the actions of Tyson, they present Tyson as part of a *decent whole*, and they present Tyson *benignly*. These themes may be thought of as elements of Tyson's projected persona. The three themes are not mutually exclusive; quite a bit of overlap exists among them, most notably between the representation of Tyson as part of a decent whole and the representation of Tyson as a benign entity; thus, those themes are presented separately in the next chapter. This chapter will focus on how Tyson disguises their actions through their discourses.

J.L. Schally, *Legitimizing Corporate Harm*, Palgrave Studies in Green Criminology, https://doi.org/10.1007/978-3-319-67879-5_5

DISGUISING ACTIONS

Throughout my analysis, I discovered that Tyson consistently utilized text and images that served to disguise their actions. Verb processes in two passages from the "Animal Well-Being" page are illustrative. Here is the first of the two passages in its entirety:

> Why animal well-being is important
>   One of our Core Values is to "serve as stewards of the animals, land and environment entrusted to us." Taking proper care of animals—treating them responsibly and with respect—is the right thing to do. It also makes great business sense. The Tyson Foods *FarmCheck*™ animal well-being program includes third-party on-farm audits, an advisory panel of animal well-being experts from around the country, and support of research on improving animal live production. (Tyson 2013a)

Examining process types allows us to see how actions get represented. Table 5.1 presents Halliday's (1976) process types for reference.

In the first sentence from this Tyson excerpt, Tyson, who would serve as steward, has agency, whereas animals, the land, and the environment are the recipients of the process. Although this is a process with a clear agent, the process is a relational one that actually has to do with qualities of the "Core Values." That is, it has no material implications. In the second part of the first sentence, Tyson becomes the recipient of the "entrusted to" process, yet this behavioral process has no agent. We are not told just who

Table 5.1   Process types

Process type	Description	Example(s)
Material	Process with material, observable consequence	"He built the house"
Verbal	Someone says something	"He told us to be quiet"
Mental	Someone thinks something	"I understand the problem"
Behavioral	Mix between mental and material	"She tasted the cake"
Relational	Process about states of being; things exist in relation to other things. Also can signify possession	"The house is blue"; "I have a pet cat"
Existential	Something exists or happens; no clear agents	"The sun came up this morning"

Source: Halliday (1976)

it is that is doing the entrusting, which leaves the reader to fill in the blank. Some possible ways that a reader might interpret this missing agent is to imagine that the agent is God,[1] especially considering that the language of the rest of the sentence conjures the idea of human dominion over land and animals put forth in the Bible:

> And God said, Let us make man in our image, after our likeness: and let them have dominion over the fish of the sea, and over the fowl of the air, and over the cattle, and over all the earth, and over every creeping thing that creepeth upon the earth. (Genesis 1:26, King James Version)

Readers may also interpret the missing agent to be societies at large that presumably consent to the practices of Tyson.

The second sentence, "Taking proper care of animals—treating them responsibly and with respect—is the right thing to do," contains several processes. First, we have "taking proper care," then "treating them," and, finally, "is." The first two processes are behavioral processes without material consequences. "Is" represents a relational process. Again we find missing agents with the first two processes. The agent, it is implied, is Tyson, who is taking proper care of animals. An alternative sentence construction that would have placed Tyson as the agent would be: "Tyson takes proper care of animals by treating them responsibly and with respect because it is the right thing to do." This alternative construction suggests that there is some concealment of action. The way the sentence is actually structured, however, the processes fall more to the mental side (recall that behavioral processes are a cross between mental and material processes), whereas my alternative construction would fall more to the material side. One gets the impression of an *idea* of "doing what is right" as much as any actual "doing." The relational process "is" attributes the quality of being "the right thing to do" to "the proper care of animals" but, again, tells us more about what Tyson thinks or believes rather than what they may actually be doing.

The third sentence, "It also makes great business sense," is a relational process where "it," referring back to the proper care of animals, is characterized as being good for business. This process also highlights an *idea* rather than an *actual practice* of Tyson. An alternative construction where Tyson is constructed as an agent involved in a material process would be, "Tyson takes care of animals because it makes good business sense." Furthermore, there is no rationale for why the proper care of animals makes good business sense. What is not being said is "we view our animals as inventory and if inventory is not looked after, we will lose money." By *not* making such an overt statement, Tyson has avoided constructing animals as being equivalent to profits, which would not necessarily sit well with a public that cares about animals, at least in the abstract. What is not said here is more telling than what is said.

Tyson continues to obscure their actions in the last sentence of the passage, "The Tyson Foods *FarmCheck*™ animal well-being program includes third-party on-farm audits, an advisory panel of animal well-being experts from around the country, and support of research on improving animal live production." There are two processes here: first, we have a relational process where the *FarmCheck*™ program is cited as including audits, experts, and research support, and second, we have a behavioral process that leans more toward a mental process where there is "support of research." Again, no explanations are given for what is materially being done by the auditors or the experts. The *FarmCheck*™ program is the agent "supporting research," but this construction does not allow explanation of what that support actually is. An alternative construction could add an additional material process, "The *FarmCheck*™ program supports research on improving animal live production by donating money to various research entities," in order to better inform readers of what is actually being done. But Tyson avoids this construction, keeping the descriptions of their actions vague and abstract.

Tyson consistently represents their actions abstractly and in nonmaterial ways, especially when representing their actions toward animals. Another example from the "Animal Well-Being" page follows:

> Putting our Words into Action
>   Whether in our plants or on the farm, we are constantly looking for new and innovative ways to better serve the farmers who supply us poultry and live-stock, as well as our customers and consumers. We've had a corporate office of animal well-being since 2000 and our commitment to responsible stewardship of the animals entrusted to us is part of our company's Core Values—to which all 115,000 of our Team Members subscribe. (Tyson 2013a)

Given the heading "putting in our words into action," it makes sense to do a process-type analysis to determine exactly how Tyson is representing their actions here. In the first sentence, "Whether in our plants or on the farm, we are constantly looking for new and innovative ways to better serve the farmers who supply us poultry and livestock, as well as our customers and consumers," the main process is a behavioral one. Tyson tells us that they are looking for new and innovative ways, but there is no elaboration on how they are looking for these new and innovative ways or what constitutes new and innovative ways. Furthermore, they are looking for new and innovative ways to better serve suppliers, customers, and consumers—*not animals*. The way that this sentence is constructed, coupled with the fact that it appears on the "Animal Well-Being" page, presupposes that new and innovative ways that would benefit suppliers, customers, and consumers would also be beneficial to animals, but this is not necessarily true. The second sentence of the passage, "We've had a corporate office of animal well-being since 2000 and our commitment to responsible stewardship of the animals entrusted to us is part of our company's Core Values—to which all 115,000 of our Team Members subscribe," like the rest of the text analyzed on the page, lacks material processes. We have a relational process where Tyson has a corporate office of animal well-being, but nothing is said of what this office actually does to ensure that animals are not suffering. There is also a behavioral process where all of the team members subscribe to Tyson's commitment to responsible stewardship. This process is quite vapid and does not communicate anything about how Tyson is actually taking care of the animals that are mysteriously entrusted to them.

Another example from the "Animal Well-Being" page is a quotation from Donnie Smith, Tyson's president and CEO: "Here's what I want people to know: at Tyson, we care enough to check on the farm; and we're determined to find better, smarter ways to care for and raise healthy animals" (Tyson 2013a). This statement contains three processes, all of which can be classified as mental: Donnie Smith "wants" people to know; Tyson "cares;" and Tyson is "determined." Again, we find no material processes that would tell readers what exactly it is that Tyson is doing.

Unexamined, the passages from the animal well-being page give the impression that Tyson is proactive and transparent in terms of animal care. On closer inspection, we find that Tyson says a lot about their *belief* in proper animal care and treatment, as evidenced by their disproportionate use of relational and behavioral processes, but very little about what they are actually doing, as evidenced by the absence of material processes. In this way, Tyson is disguising their actions while simultaneously giving the appearance of being forthcoming, by the way the text is constructed.

My analysis of passages from the "Environment" page yielded slightly different results from the "Animal Well-Being" page, but overall the results still provide evidence of concealing action. On the surface, it would seem that Tyson is more transparent about their impact on the environment than they are about the treatment of animals on their farms and in their processing facilities. Tyson offers the following, relatively straightforward, statement on conserving water:

> Water conservation has been an important area of focus for us for many years. We employ programs and technologies to conserve this natural resource. Our first priority is to ensure the safety of our food products, and we will never reduce water usage in situations where food safety could be compromised. We are pleased with the water conservation efforts made and the results we have achieved so far. Our water conservation efforts, along with several facility closures, have led to a water usage reduction of 10.9 percent since October 2004. (Tyson 2013b)

Although this passage does not contain any material processes, Tyson provides us with some actual data about their water conservation efforts which points to transparency—*not* disguising actions. Tyson does not offer, however, any context for the 10.9 percent drop in water usage: one might reasonably wonder what the rates of consumption were before the drop. Tyson also demonstrates slightly greater transparency with respect to their actions in their discussion of their carbon footprint:

> We've been tracking, calculating and publicly reporting our greenhouse gas (GHG) emissions since 2004. We continue working with the U.S. EPA regarding GHG inventory information related to mandatory GHG reporting requirements. We've made important strides in the areas of energy efficiency, fuel consumption and renewables. (Tyson 2013b)

Here, Tyson tracks, calculates, and reports—all material processes. Many actions are presented abstractly throughout the previous passages, however. For example, the second sentence of the first passage, "We employ programs and technologies to conserve this natural resource," is not very informative and leaves the reader to speculate as to what exactly the programs and technologies are. The last sentence of the passage about Tyson's carbon footprint states that they have made "important strides," but readers learn nothing about what these strides are. Other excerpts from the

"Environment" page also present actions abstractly. For example, Tyson's environmental policy states, "with our Core Values as our foundation, it is our policy to conduct business in a safe, responsible manner with respect for the environment and for the well-being of our Team Members, customers, and neighboring communities" (Tyson 2013b). The process of "conducting of business in a safe, responsible manner" is quite nonspecific; we are not told what this actually means. Likewise, Tyson's statement about their Poultry Environmental Stewardship Award is generalized: "Tyson Foods depends on independent Family Farmers to supply livestock and poultry. In an effort to recognize the farmers who support us, Tyson Foods maintains a Poultry Environmental Stewardship Awards program." Although Tyson is more forthcoming about environmental practices than they are about animal welfare practices, Tyson's actions are still obscured in the text.

Also notable is the glaring absence of photographs or images on the "Animal Well-Being" page. All of the other pages I examined contain at least one photograph or other image. Indeed, most of the pages on Tyson's site contain several photographs or images. The absence of photographs or images on the "Animal Well-Being" page is therefore quite conspicuous. It is easy to interpret this absence of photographs and other images as indicative of Tyson trying to conceal the conditions in which their animals are raised and processed. Further supporting this interpretation is the use of cartoons as opposed to real photographs on the "Making of a Meal" page, which is supposed to depict all of the stages of processing from farm to table. One such cartoon depicts a feedlot covered in grass where cows have ample space and seem to be enjoying life. This cartoon is coupled with the following passage: "We have relationships with feedlots that finish raising the cattle we buy and with livestock operations that raise the hogs we buy to make sure they are healthy, treated with respect, cared for properly, and grown to the appropriate weight and age" (Tyson 2013c). The next cartoon shows a rendition of a processing plant. The processing plant is pictured as clean and bright; the workers stand at an assembly line where packages pass by on a conveyer belt. The processing plant cartoon is coupled with the following relatively vapid passage:

> More than 90,000 Team Members work at our plants in the U.S, to make the thousands of beef, chicken, pork, and prepared foods products we produce. Did you know that we make more pepperoni and other pizza toppings than anyone else in the U.S. and are the second largest producer of corn and flour tortillas?

By using cartoons, Tyson avoids showing the public the actual conditions of their farms and processing plants, and by sequencing the processing plant cartoon right after the feedlot cartoon, Tyson distracts us from the slaughter that takes place between the feedlot and the packaging of the meat.

Actions are also disguised through metonymy—using a related term to refer to something instead of the actual name. Metonymy is common practice in general within discourses of meat-eating; it helps people avoid the idea that they are consuming the flesh of dead animals. Foer notes that meat-eating has an "invisible quality," which he tries to make less invisible by offering us a recipe for "Stewed Dog, Wedding Style" (2009: 28). The use of metonymy contributes to the invisible quality pointed out by Foer (2009). Some examples of Tyson's use of metonymy include consistently using "poultry" to talk about chickens. By referring to chickens as poultry, chickens are rendered inanimate. Other euphemisms used for chickens include "breeder," "ice-packed broiler," and specific product names such as "Chick'n Quick." Tyson uses common terms that conceal the origin of meat, such as "beef" and "pork," and consistently uses the term "protein" to collectively refer to the flesh of dead animals. Stibbe notes that we refer to animals differently after they die: "It is possible to say some chicken, some lamb, or some chicken leg, but some human and some human leg are ungrammatical" (2001: 151). Thus, by employing the word "protein," Tyson perpetuates the cultural disguising of meat-eating.

As Tyson disguises actions, they simultaneously present themselves as being wholesome and good. The next chapter will focus on the ways that Tyson achieves this identity through their discourse.

NOTE

1. When I reference God or religion throughout the book, I have in mind God and religion in the Judeo-Christian tradition, which is dominant in the United States.

REFERENCES

Foer, Jonathan S. 2009. *Eating Animals*. New York: Little Brown and Company.
"Genesis 1:26 (King James Version)." Bible Gateway. http://www.biblegateway.com/passage/?search=Genesis+1%3A26&version=NKJV. Accessed 23 Mar 2010.

Halliday, Michael A.K. 1976. Types of Process. In *Halliday: System and Function in Language*, ed. Gunther R. Cress, 159–173. Oxford: Oxford University Press.

Stibbe, Arran. 2001. Language, Power and the Social Construction of Animals. *Society & Animals* 9: 145–161.

Tyson Foods. 2013a. *Animal Well-Being*. http://www.tysonfoods.com/Ways-We-Care/Animal-Well-Being.aspx. Accessed 11 Aug 2013.

———. 2013b. *Environment*. http://www.tysonfoods.com/Ways-We-Care/Environment.aspx. Accessed 11 Aug 2013.

———. 2013c. *Heritage*. http://www.tysonfoods.com/OurStory/Heritage.aspx. Accessed 11 Aug 2013.

CHAPTER 6

Being Good: Or at Least Not Bad

Abstract This chapter discusses the ways that Tyson presents themselves benignly and as part of a decent whole. Tyson presents themselves as part of a decent whole by aligning themselves with popular cultural values, and they present themselves benignly by constructing an image of a good corporate citizen/neighbor.

Keywords Narrative • Metaphor • Presupposition

Tyson presents themselves as part of a decent whole by aligning themselves with popular cultural values, and they present themselves benignly by constructing an image of a good corporate citizen/neighbor. This chapter deconstructs how Tyson hides behind wholesomeness.

Part of a Decent Whole

Tyson presents themselves as part of a decent whole by drawing on and aligning with popular American values and motifs, such as a strong work ethic based in a capitalist economy. In so doing, Tyson draws the reader in and gains his/her approval, for they presumably share in that popular culture. The theme is most evident through Tyson's origin story, which is outlined on the "Heritage" page (see http://www.tysonfoods.com/our-story/heritage). Tyson's origin story contains all of the elements of a fully formed narrative identified by Labov (1972) as it recounts the founding of Tyson, its early years, and its evolution into the company it is today. In this narrative, Tyson presents themselves as good, wholesome, and as part of traditions that are worth preserving. They preface their story with the following passage:

> Few modern food companies have real connections to their pasts, but these links remain strong within our culture. Our company endures because this culture endures. It's part of who we are today and who we will be tomorrow. (Tyson 2013c)

Tyson evokes the idea of tradition and having strong roots throughout the origin story narrative. For example, when Tyson tells us that they moved into new corporate offices in the early 1970s, they tell us that they are still in the same place today. Tyson's official "core values" were created in 2002, but when relating this development the website states that they were created to "reflect the time-honored principles our Team Members have lived by since the early days of Tyson Foods" (Tyson Foods 2013c). Tyson relates the significance of the company's history in a way that echoes American patriotism and pride in US history.

In addition to drawing on the narrative of significant traditions, Tyson's origin story also utilizes the rags-to-riches formula to position themselves as a company started by ordinary folks. Tyson tells the tale of humble beginnings overcome through hard work and dedication. The following excerpt is illustrative: "The Tyson Foods Story begins during one of the most difficult periods of American history—the Great Depression. In 1931, John W. Tyson moves his wife and small son to Springdale, Arkansas, in search of new opportunities" (Tyson 2013c). This passage is coupled with a sepia-toned photograph of a run-down shack with no caption. It is unclear if the structure is supposed to represent the first chicken house of Tyson Foods, the living quarters of John and his wife, or something else. Regardless, the photo of the shack emphasizes the poor conditions at the time that the company started; our attention is drawn to the dilapidated conditions, possibly evoking empathy for John Tyson as he struggled to make his way in the world with his young family.

Tyson also uses religious references, both implicit and explicit. The use of religion allows Tyson to appeal to a large segment of the US population.[1] Tyson implicitly invokes God when they state that they are "stewards of the animals, land, and environment entrusted to us," and explicitly when they state "we strive to be a faith friendly company," and "we strive to honor God and be respectful of each other, our customers, and other stakeholders" (Tyson 2013a). By conjuring notions of religiosity, Tyson constructs themselves as good and pure.

Other examples of Tyson positioning themselves in line with larger cultural narratives abound. This quoted recollection from Don Tyson, son of founder John Tyson, is demonstrative: "I left the university [University of Arkansas] in 1952, and from that day until 1963, the year I took the company public, I worked in the business six days a week and on Dad's farm on the seventh day" (Tyson 2013c). The quotation supports the populist sentiment of anti-intellectualism representing Don Tyson as a champion of "real," practical work. Here, Tyson aligns themselves with the larger cultural narratives concerning entrepreneurship and meritocracy.

Indeed, Tyson narrates a classic Horatio Alger tale in which the company founders (John and Don Tyson) are cast as rugged individuals who pulled themselves up by their bootstraps. Through this story, Tyson conveys a message that they are not so different from the average hard-working American and that they are deserving of their success. A photograph of John and Don Tyson on the "Heritage" page supports this interpretation. The black-and-white photograph pictures John and Don Tyson seated at a desk in a sparsely decorated office that communicates the idea that these men were not frivolous. Likewise, the men are not wearing business attire, but rather the uniform of the average worker, which gives the impression that these are ordinary, everyday men. In the background of the photo hangs a plaque, which suggests their having been honored for their work, but the fact that the plaque is in the background of the photo also suggests humility.

Humility—being just like your everyday, average American—is a consistent theme throughout Tyson's origin story, from the company's founding to recent years. Even those at the top-most rungs are characterized as humble. CEO Leland Tollett (named CEO in 1991) is pictured in the uniform of an average blue-collar Tyson plant worker, serving up platters of food, despite the fact that he was then head of a company with over $3.9 billion in net annual sales (*The New York Times* 1992). The image suggests that even as Tyson became a highly successful corporation, they never lost sight of where they came from.

Throughout the origin story, the founders and leaders of Tyson are consistently cast as savvy, but genuine and caring businessmen—exemplars of industriousness and trustworthiness. They have the earthy common sense of a trusted advisor. Following are some of the evaluations (all from Tyson 2013c):

- "John says later, 'I decided early that, if you had the best chicks in the area you'd have the best customers and get the best results.'"
- "As Don would often say, 'Our people are the heart of our company.'"
- "'Don had an uncanny ability to acquire the right company at the right time,' Leland would say later. 'But the real success of the company is the result of a genuine commitment by Don to always take care of our people.'"
- "In 1979, Don writes in our annual report: 'The modern organization exists to provide a specific service to society. For Tyson Foods, that means high-quality poultry and other food products. But the corporation has to be in society, to be a good neighbor, and to do its work within a social setting. ... Tyson Foods believes that, if we don't take an active part in the community, we won't deserve a place in it very long.'"

By presenting themselves in this shrewd yet down-to-earth way, Tyson positions themselves to win over the hearts and minds of a public that values both cleverness and unpretentiousness.

Analysis of discursive presuppositions—or assumptions embedded in language—yielded still more evidence of Tyson aligning themselves with the American ethos within the origin story. The following quotation about earning profits is informative: "We strive to earn consistent and satisfactory profits for our shareholders and to invest in our people, products, and processes." This statement presupposes that pursuit of profit is good and the right thing to do, but more importantly, it presupposes that readers will agree. Indeed, most people probably would agree, allowing Tyson to capitalize on that shared cultural belief. Also found throughout the origin story and other pages in the sample are presuppositions about the legitimacy of meat-eating. For example, the origin story refers to new "poultry products" being introduced throughout the years under the heading of "Convenient Chicken for Everybody," which presupposes that eating chicken is normal. Furthermore, one of Tyson's core values is "We feed our families, the nation, and the world with trusted food products" (Tyson 2013a); again, meat-eating is taken for granted in this statement that casts Tyson's products (i.e., meat) as something that is permissible and normative. The "Making of a Meal" page has Tyson repeatedly referring to raising and processing animals for food in ways that assume that these practices are of the natural order. For example, in reference to their supply partners, Tyson states, "We have relationships with feedlots that finish raising the cattle we buy and with livestock operations that raise the hogs we buy to make sure that they are healthy, treated with respect, cared for properly, and grown to the appropriate weight and age"(Tyson 2013f). Given that the United States consumes more meat per capita than any other country (ChartsBin 2013), I would argue that eating meat has assumed the status of a cultural value (see Adams 1990). Therefore, presuppositions about meat-eating as a normal and unproblematic practice serve to legitimize the business of Tyson, as well as perpetuate the "invisible quality" of eating animals (Foer 2009: 29).

In sum, Tyson presents themselves as part of a decent whole by communicating that they are not so different from the everyday average American—that they share the same values, goals, and ideas as everyone else. Tyson as part of a decent whole is exemplified by their telling readers that they are our friends and neighbors: "we volunteer, shop, vote and raise our families in communities across America" (Tyson Foods 2013d). In the next section, I explore how Tyson presents themselves in a benign way—that is, how Tyson represents themselves as good corporate citizens. Similar to the discursive moves just examined, Tyson presents themselves as decent and wholesome, but in ways that do not necessarily reflect larger cultural values and ideas.

BENIGN PRESENTATION OF SELF

I found throughout the pages examined that Tyson presents themselves and their products in a very wholesome, benign way. As they tell it, Tyson is friendly, honest, and socially responsible. The statement "Our people are the heart of our company" (Tyson 2013b) utilizes the metaphor of the heart to drive home the point that Tyson believes that their employees are the most vital part of their business. But the heart also invokes thoughts of love, care, and compassion. The metaphor of the heart is used again on the "Relationships" page: "Farmers who work the land and care for the animals are at the heart of Tyson Foods' success" (Tyson 2013d).

Tyson consistently represents their products as wholesome and good. Although it is not remarkable that Tyson would present their products in such a way, they seem to suggest that consumers who use their products will also become wholesome and good. On the "Making of a Meal" page (see http://www.tysonfoods.com/loved-brands/the-making-of-a-meal), Tyson states that "[a]t the end of the day, our Team Members work together with Family Farmers, supplier partners, retailers, restaurateurs, and other foodservice partners to help you create the special meal-time memories your family deserves" (Tyson 2013f). The previous quotation is coupled with a cartoon image depicting three Polaroid pictures on a picnic table. Two of the cartoon photographs are foregrounded: one shows a family eating a meal together and the other shows a group of people outdoors enjoying a cookout. These images invoke nostalgia and thoughts of good times and special occasions.

Tyson promotes the idea that their products facilitate the making of good memories with family. They use the word nourish metaphorically to underscore that Tyson products not only provide sustenance, but also foster family togetherness.

> Even though sitting down together for a nightly meal doesn't happen in every household, we still believe meal time is prime time for memory making. Whether you're on the go, or around the table with the ones you care about most, good times start with great food. Working together to produce food you want and trust creates lasting relationships among our team at Tyson Foods—much the same way your relationships might be nourished at meal time. (Tyson 2013d)

Benign presentation of self is also evident through Tyson's use of cartoons throughout the "Making of a Meal" page. By using cartoons, Tyson brings a whimsy to their website that helps portray the company as light-hearted, friendly, and safe. Cartoons have been used in advertising and other propaganda throughout history to appeal to kids or to diminish ideas of certain products as harmful (e.g., the Joe Camel character used in the marketing of Camel cigarettes).

By presenting actions in an abstract manner, Tyson easily constructs itself as compassionate. "Making Great Food. Making a Difference™" is the "purpose statement" of Tyson Foods (Tyson 2013c). The action in this statement, making a difference, is abstract: the statement provides no real information about how exactly Tyson is making a difference, but the statement implies that whatever it is, it is good. Looking more closely at what Tyson says about how they make a difference, they highlight their commitment to diversity, "we strive to be a company of diverse people" (Tyson 2013a); their role in providing for our nation's children, "9 new products in 2012 that meet updated USDA school lunch standards" (Tyson 2013e); and their charitable donations, "18 million pounds of protein donated to hunger relief agencies since 2010" (Tyson 2013e). Thus, we see a mix of abstract and more concrete statements but even the concrete statements lack context. For example, donating 18 million pounds of protein within a three-year time frame on the surface may seem impressive until one compares that figure to the fact that just *one* food bank in New York City distributes more than 74 million pounds of food each year (Food Bank for New York City 2014) as part of a national system of regional food banks that distribute more than 3 billion pounds of food annually (Feeding America 2014). So while the information about their donations appears straightforward, the larger framework of national food donations is concealed.

Continuing in the vein of presenting themselves as a compassionate, good corporate citizen, the sole photograph on the "Environment" page shows a Tyson worker bending to take a water sample from a small stream with farm buildings in the background. The setting of the photograph is idyllic: fair weather, clean water, and green grass. This image communicates that Tyson is following the rules and putting the safety of the environment and human health as a primary concern. More broadly, the photograph is symbolic of self-regulation and Tyson's commitment to this practice and principle.

Tyson also constructs themselves in a benign way through their use of neutral quoting verbs. The use of certain quoting verbs can have the effect of persuading the reader to think a certain way about the speaker and the validity of what the speaker is saying. The web pages in the sample contain very few quoting verbs, in fact only in the "Heritage" section where the company's origin story is recounted. Here, where quoting verbs do appear, it is most often with a neutral construction. A few examples follow with quoting verbs in boldface:

- John **says** later, "I decided early that, if you had the best chicks in the area, you'd have the best customers and get the best results."
- As Don would often **say**, "Our people are the heart of our company."
- "Don had an uncanny ability to acquire the right company at the right time," Leland would **say** later.
- Don **writes** in our annual report: "The modern organization exists to provide a specific service to society."

- "We **say** it in three words: segment, concentrate, and dominate. We find something we think we can do, focus on it, and then aim to be number one at it. Most of the product categories we dominate are things we started," **says** Don in the 1986 annual report.

The use of neutral constructions may connote even-handedness or practicality more so than other quoting verbs such as "claimed," "explained," or "remarked." Thus, the use of neutral constructions may serve the purpose of presenting the speaker as benign. The sole exception to the use of neutral quoting verb is instructive: "Donnie Smith **announces** a new purpose statement to reflect our rich heritage of *Making Great Food. Making a Difference.*™" The quoting verb "announces" connotes assertiveness and sounds official and formal leading to an interpretation of legitimacy of the statement and the speaker.

In sum, Tyson constructs an image of wholesomeness that promotes trust in them and their products. Tyson is a good corporate citizen who cares about us. Through analysis of its website, I found that Tyson distances themselves from harm-doing (disguising actions), while idealizing its image (part of a decent whole and benign presentation of self). That is, Tyson achieves distance from the ideological "bad" and proximity to the ideological "good." Smith (2005) notes that such binary codes are "responsible for classifying the world and so doing according to moral criteria, detailing the qualities and attributes of the sacred and profane, polluted and pure" (14). This binary of cultural distancing and alignment permits Tyson's continued success.

NOTE

1. A recent Harris Poll revealed that although belief in God, miracles, and heaven has declined, 74 percent of Americans believe in God and approximately 59 percent of Americans identified themselves as "very religious" or "somewhat religious" (Harris 2013).

REFERENCES

Adams, Carol J. 1990. *The Sexual Politics of Meat*. New York: Continuum.
ChartsBin. 2013. *Current Worldwide Annual Meat Consumption Per Capita*. http://chartsbin.com/view/bhy. Accessed 16 Feb 2014.

Feeding America. 2014. *Our Food Bank Network.* http://feedingamerica.org/how-we-fight-hunger/our-food-bank-network.aspx. Accessed 14 June 2014.

Foer, Jonathan S. 2009. *Eating Animals.* New York: Little Brown and Company.

Food Bank for New York City. 2014. *Food Sourcing and Distribution.* http://www.foodbanknyc.org/our-programs/food-sourcing-and-distribution.   Accessed 14 June 2014.

Harris. 2013. *Americans' Belief in God, Miracles and Heaven Declines.* http://www.harrisinteractive.com/NewsRoom/HarrisPolls/tabid/447/ctl/ReadCustom%20Default/mid/1508/ArticleId/1353/Default.aspx. Accessed 13 June 2014.

Labov, William. 1972. *Sociolinguistic Patterns.* Philadelphia: University of Pennsylvania Press.

Smith, Philip. 2005. *Why War? The Cultural Logic of Iraq, the Gulf War, and Suez.* Chicago: University of Chicago Press.

The New York Times. 1992. Company News; Tyson Plans to Expand into Seafood. *The New York Times,* June 17. http://www.nytimes.com/1992/06/17/business/company-news-tyson-plans-to-expand-into-seafood.html. Accessed 14 June 2014.

Tyson Foods. 2013a. *Core Values.* http://www.tysonfoods.com/Our-Story/Core-Values.aspx. Accessed 11 Aug 2013.

———. 2013b. *Fiscal 2013 Fact Book.* http://ir.tyson.com/files/doc_downloads/Tyson%202013%20Fact%20Book.pdf. Accessed 17 Apr 2014.

———. 2013c. *Heritage.* http://www.tysonfoods.com/OurStory/Heritage.aspx. Accessed 11 Aug 2013.

———. 2013d. *Relationships.* http://www.tysonfoods.com/Our-Story/Relationships.aspx. Accessed 11 Aug 2013.

———. 2013e. *Sustainability Report.* http://www.tysonfoods.com/Ways-We-Care/Sustainability-Report.aspx. Accessed 11 Aug 2013.

———. 2013f. *The Making of a Meal.* http://www.tysonfoods.com/Our-Story/The-Making-of-a-Meal.aspx. Accessed 11 Aug 2013.

CHAPTER 7

Taking Stock, Taking Action

Abstract This chapter provides a synthesis and reflection on the preceding chapters. In sum, my project finds that Tyson obscures their actions toward animals and the environment by talking about them abstractly or not at all. The chapter also reflects on possible future research as well as how change may be effected.

Keywords Discourse • Green criminology • Social harm

What did I do, what did I find, and why does it matter? These are the three questions that concern me in this chapter. In addition to answering them, I will offer suggestions for future research in the space where cultural criminology, discourse analysis, and green criminology meet.

What Did I Do?

I conducted a critical discourse analysis with an eye toward nominalizations, process types, narrative, metonymy, metaphor, and abstractions. In addition, my analysis was concerned with the underlying meaning of images, which I uncovered and explored by asking a "sequence of specific questions" about the images (Machin and Mayr 2012: 49). I sought the specific discourses that facilitate the cultural legitimation of harm that is parcel to industrial agriculture. As such, I scrutinized those web pages that were most relevant to my research questions: How does Tyson construct themselves and their actions toward animals and the environment to the public? What do these constructions accomplish? The following seven web pages were examined:

1. Animal Well-Being: Tyson tells about their commitment to treating animals properly and promotes their *FarmCheck*™ program.
2. Environment: Tyson tells about conducting business in a way that respects the environment and highlights the reduction of water consumption.
3. Core Values: Tyson tells us who they are, what they do, and how they do it.
4. Heritage: Tyson tells their origin story and the evolution of the company.
5. Sustainability Report: Tyson uses an infographic to highlight some of the main information from their larger report.
6. The Making of Meal: Tyson follows their product(s) from the farm to the table.
7. Relationships: Tyson tells about their relationships with various partners—contract farmers, other livestock suppliers, community partners, and others.

These pages, among many others on the website, offered the most promise for information regarding my three main areas of interest: Tyson's actions toward animals, Tyson's actions toward the environment, and Tyson's presentation of self.

WHAT DID I FIND?

First, as this project was inspired by Stibbe's (2001) discourse analysis of texts produced by animal product industries, it is worthwhile to consider how our findings are similar and how they are different. Similar to Stibbe (2001), I found the ample use of metonymy throughout Tyson's website. In Stibbe's work, the discourses examined consisted mainly of texts meant for industry insiders. Part of what those discourses accomplished was to allow the insiders to view their actions toward animals as actions toward objects. The discourses on Tyson's website permit the average American to see his/her own actions toward animals (eating them) as nonproblematic.

Through my analysis, I demonstrated that Tyson Foods discursively constructs an identity of a good corporate citizen while simultaneously disguising their harmful actions. My findings have also demonstrated that the absence of certain discourses permits the cultural legitimation of harm. On the matter of the harm they do to animals, Tyson is abstract at best. In fact, they are mostly silent. Tyson tells us little more than that they are committed to animal well-being and that it is important. Specifics of what this commitment means to Tyson and how they implement it in their business workings are not provided. Also absent from the pages I examined were photographs of any Tyson farms or Tyson farm animals; the farms and farm animals were represented only through cartoons. The use of cartoons and absence of photographs is conspicuous because when compared to the rest of the pages analyzed—indeed, the rest of the website—we find that cartoons are not used much and that photographs abound.

Scrutinizing what is not said is especially important in the analysis of harm to animals. As Presser and Schally explain, "the work of exclusion is perhaps the most harm-conducive stratagem of all....the modern-day machinery whereby animals are raised and slaughtered for their meat and other bodily products—the primary means by which animals are made to suffer in contemporary Western societies—runs on our dissociation and silence" (2013: 181–182). Activists have long noted that silence is consent, at least among those who have the freedom/ability to speak. It is not surprising that Tyson excludes discourses on how animals get treated because this exclusion allows people to remain willfully ignorant of the suffering endured by the animals they will eat. As long as people avoid knowing what is really going on, they can continue having bacon without much, or any, cognitive dissonance. Tyson's discourse is complicit in this avoidance of knowledge.

Tyson is more transparent and provides more detail when talking about environmental issues as compared to their statements on animal welfare. This disparity may be attributable to the fact that there are more laws regulating pollution than there are regulations for the treatment of farm animals. How animals are treated on the industrial farm seems to be mystified to a much greater degree than how the ecological environment is treated. It could be argued that Tyson is only more forthcoming with environmental issues because they *have* to be and so long as laws regarding farm animal welfare are nonexistent, they will continue to present an image of care and concern for animals that, when examined closely, falls apart. At the same time, even though, on the surface, it appears that Tyson is more transparent about their environmental record, there is still heavy use of abstraction in how they talk about their relationship to the environment. They give us some specifics, such as how much they have reduced their water consumption, but this disclosure is without context. The main points on the pages where they talk about the environment have, much like their talk about animals, to do with commitments and beliefs—not actions.

In terms of how they construct their image, Tyson tells feel-good stories that present them as a good corporate citizen. With that in mind, it is worth pointing out that corporations may not legally present blatantly false claims.[1] As such, the identity work of corporations involves more subtle maneuvers that allow corporations to construct images that are positive or at least neutral. It is difficult to say with any certainty, however, that the construction of a positive image is the *intention* of the people behind Tyson's website, although what we know about public relations or "spin" seems to suggest as much. The tools of CDA allow one to speak to the *results* of identity work only and not the motives of communicators. Indeed, CDA is generally not concerned with the motives of those who would create discourse; rather, it is concerned with what the discourse has accomplished regardless of the intention. For example, Wood and Kroger (2000) note that although evidence of intention may be lacking in an instance of nominalization that obscures agency, we can still say that the use of nominalization obscures agency because that is what nominalization does grammatically; whether the writer or speaker purposely used nominalization is of no significance. Regardless, the purpose of this project was not to necessarily argue that Tyson has *intended* to dupe people into believing things that are not true. The purpose of the study was to examine the discourses used by Tyson and what those discourses accomplish.

The stories told on the website—stories of humble beginnings and hard work building to success—reinforce dominant normative logics of capitalism. By drawing on and aligning themselves with larger cultural themes such as individualism and meritocracy, Tyson's practices seem unproblematic. Tyson's discourses, particularly the origin story, consistently construct them as "good," allowing them to benefit from the binary codes of good and evil—because if you are one, then you surely cannot be the other. Narrative criminologists hold that stories provide the impetus for action (Presser 2009). What I find here is that Tyson's origin story may not necessarily be an instigator of action in the first instance, but rather allows the harm inflicted by their practices to continue. So long as the public absorbs their story and message—of being part of a decent whole and a benign presence in society—then their harmful actions can be tolerated.

WHY DOES IT MATTER?

This project is significant in several ways. First, it brings a focus to the role of language and other symbols in the perpetuation of harm by utilizing a critical discourse method. In addition, because of this project's concern with the cultural legitimation of harm and how this is achieved through mediated messages on a corporate website, this work stands as an empirical example of the merging of green and cultural criminology. Recall that cultural criminology is about power and how power is summoned within the culture of the late-modern era. The merging of cultural and green criminology is an important intersection because the patterns of power and oppression that play out regarding animals and the environment are not discernable without considering the cultural context of late modernity.

This project adds to the social harm literature by taking to task an institutionalized practice that is often not labeled as harmful even within some of the most comprehensive definitions.[2] Note David Wästerfors' recent critique of Presser's (2013) book, *Why We Harm?*, where he takes issue with her categorization of animal-killing/meat-eating (among other actions) as harm, stating that "there is no cultural consensus today that killing and eating non-human animals...are harmful acts" (2014: 271) and labeling as "utopian" her treatment of those activities as such. Even critical sociologists seem to have a blind spot where nonhuman well-being is concerned. Indeed, the social harm perspective (see Hillyard et al. 2004) is, though more inclusive of what can and should be considered harm than mainstream criminology, still quite anthropocentric. That is, who can get constructed as "victim" is limited to human beings. Nonhumans are generally not considered as victims in their own right. Although it may be the case that killing animals for their flesh is not widely considered to be harm *currently*, my hope for this work is to create a space for alternative cultural critique and perspectives.

DIRECTIONS FOR FUTURE RESEARCH

Fairclough (1992) observes that some CDA approaches place too little emphasis on discourses of resistance and transformation. That is, the inquiry is concerned with power abuse and harm, not with opposing these. My project of a critical discourse analysis of Tyson's website focused mainly on reproduction—on "how subjects are positioned within formations and how ideological domination is secured" (Fairclough 1992: 33–34). As such, future research should also examine discourses of resistance, which might sample from websites of animal welfare/animal rights groups or even websites of small, organic farms.[3] I am under no illusions but that these websites also "spin" what they do in particular ways. Yet, comparison with my results might lead advocates to an understanding of how—to use Presser's (2013) terms—to tell "true stories" for the sake of "unmaking misery."

Future work could utilize Internet archives to examine the discourses of Tyson's website over time and how they have changed. It would be useful to determine if changes in Tyson's website over time could be related to larger sociocultural changes, such as the increase in ethical consumerism in the United States. Future studies might include other cases to compare to determine if the discursive constructions used by Tyson are common across big agribusinesses. For example, discourses from the websites of other large agribusinesses involved in meat production, such as Smithfield, Perdue, and Pilgrim's Pride, could be examined in the same way that I have done here with Tyson, to answer the question of whether other large agribusinesses follow the binary codes of distancing from "bad" and aligning with "good" and to examine how other agribusinesses represent their actions toward animals and the environment. I suspect that I have located a general pattern but empirical verification is necessary.

In sum, my project has found that Tyson obscures their actions toward animals and the environment by talking about them abstractly or not at all. Working with this exclusion of meaningful information about animals and the environment is the underlying binary code used throughout the discourse on Tyson's website—the quintessential good/bad archetype. Together, the concealment of action and the use of binary codes allow Tyson to reproduce the legitimacy of their own actions and also the legitimacy of the cultural values they connect with and identify as good, such as capitalism, the Protestant ethic, and meat-eating.

Based on my analysis, how could Tyson be rehabilitated? Just as discourses facilitate the cultural legitimation of harm, they have the power to facilitate the delegitimation of harm. Imagine that Tyson was transparent about their practices with animals and that they connected to values that said "minimize the doing of harm." I would suggest that in such a scenario—that is, if Tyson constructed their actions differently—the possibility for acting differently exists.

NOTES

1. US federal law specifies that ads "must be truthful, not misleading, and, when appropriate, backed by scientific evidence." Enforcement of truth-in-advertising laws falls under the jurisdiction of the Federal Trade Commission with the same requirements regardless of where an ad appears (e.g., in newspapers, magazines, online) (Federal Trade Commission n.d.)
2. See my discussion of Henry and Milovanovic (1996) and Schwendinger and Schewendinger (1970) in Chap. 1.
3. Some small farms have adopted a "glass walls" approach, whereby all farm operations are observable to the public. Fair Oaks Farm is one such example (http://fofarms.com/).

REFERENCES

Fairclough, Norman. 1992. *Discourse and Social Change*. Cambridge: Polity Press.

Federal Trade Commission. n.d. *Truth in Advertising*. http://www.ftc.gov/news-events/media-resources/truth-advertising. Accessed 30 June 2014.

Henry, Stuart, and Dragan Milovanovic. 1996. *Constitutive Criminology: Beyond Postmodernism*. London: Sage.

Hillyard, Paddy, Christina Pantazis, Steve Tombs, and Dave Gordon, eds. 2004. *Beyond Criminology: Taking Harm Seriously*. London: Pluto Press.

Machin, David, and Andrea Mayr. 2012. *How to Do Critical Discourse Analysis*. Thousand Oaks: Sage.

Presser, Lois. 2009. The Narratives of Offenders. *Theoretical Criminology* 13: 177–200.

———. 2013. *Why we Harm*. New Brunswick: Rutgers University Press.

Presser, Lois, and Jennifer L. Schally. 2013. Institutionalizing Harm in Tennessee: The Right of the People to Hunt and Fish. *Journal of Sociology and Social Welfare* 40: 169–184.

Schwendinger, Herman, and Julia Schwendinger. 1970. Defenders of Order or Guardians of Human Rights? *Issues in Criminology* 5: 123–157.

Stibbe, Arran. 2001. Language, Power and the Social Construction of Animals. *Society & Animals* 9: 145–161.

Wasterfors, David. 2014. Book Review: Why We Harm. *Acta Sociologica* 57: 271–272.

Wood, Linda A., and Rolf O. Kroger. 2000. *Doing Discourse Analysis: Methods for Studying Action in Talk and Text*. Thousand Oaks: Sage.

Appendix

Research Methods or How I Digested What Tyson Was Serving Up

In order to grasp how the harms caused by Tyson Foods continue as they do, I closely examined how Tyson constructs themselves and their actions to the public. Critical discourse analysis (CDA) best served the purpose of the project as the epistemological position of the project is in line with a "hermeneutics of suspicion" or demystification in which the researcher "problematizes the participants' narrative and strives for explanation beyond the text" (Josselson 2004: 1). A hermeneutics of demystification is concerned not only with eliciting the underlying meanings in texts, but also with examining what is conspicuously absent from texts (Josselson 2004).

Design/Sample

This project is a case study of Tyson Foods and specifically the "identity work" they perform via their corporate website. As noted elsewhere in the text, I chose Tyson Foods as an exemplar of big agribusiness as it is the largest corporation involved in livestock production. Data for the project include text and images from Tyson Foods' website in 2013. Because content on websites is subject to change, I compiled all text and images from all pages on the website into a text file. I created this text file in August 2013, so the analysis is based on website content at that time. Although

© The Author(s) 2018
J.L. Schally, *Legitimizing Corporate Harm*, Palgrave Studies in Green Criminology, https://doi.org/10.1007/978-3-319-67879-5

various parts of the corporate website work together to paint a picture of the good corporate citizen, including strategic use of stories, images, and other symbols, this detailed discourse analysis focuses on specific parts of Tyson's website. I examined content (text and images) from the following pages: "Heritage," "Core Values," "The Making of a Meal," "Relationships," "Animal Well-Being," "Environment," and "Sustainability Report." I chose these specific pages because they are most relevant to the interests of this project, namely, Tyson's construction of their identity as well as how they represent their actions toward animals and the environment.

Procedures/Analysis

Guided by the principles of CDA, I scrutinized data from the website for evidence of obscuring harm, including, but not limited to, evasions of agency and denial of harm and victim. I employed some well-honed techniques used in discourse analysis concerning the construction of speech, actions, and things, and the use of metaphors and narratives.

The Theoretical Foundations of Critical Discourse Analysis

Discourse analysis is a qualitative approach that consists of various ways of thinking about and examining discourse (Wood and Kroger 2000). What counts as discourse can vary, but generally, and specifically for this project, any use of language (written or spoken) or other symbols (e.g., photographs, drawings) that communicate an idea may be considered discourse. Although discourse analysis employs some of the tools of linguistic analysis, it moves beyond such analysis by being grounded in the idea that discourses constitute social practices (Fairclough 1992; Wodak and Meyer 2009; Wood and Kroger 2000). Wodak and Meyer note that describing discourse as social practice "implies a dialectical relationship between a particular discursive event and the situation(s), institution(s), and social structure (s) that frame it" (2009: 6).

CDA is particularly concerned with the relationship between power and language (Fairclough 1992; van Dijk 1993). It shows us "how discourse is shaped by relations of power and ideologies, and the constructive effects of discourse has upon social identities, social relations and systems of knowledge and belief" (Fairclough 1992: 12). Critical discourse analysts thus scrutinize ideologies and power through systematic examination of discourse (Wodak and Meyer 2009). CDA differs from discourse analysis per se in that critical approaches look at how discourses get used to sustain power relations and/or harmful relations, whereas noncritical approaches do not necessarily account for the role of power and ideology in producing discourse. Noncritical approaches tend to assume that a shared understanding exists among those engaging in discourse: participants share a frame of reference and thus interpret events similarly. For example, a noncritical analysis of discourse in a classroom setting is likely to operate on the assumption that the relationship and power dynamics in this setting exist as they should, naturalizing "dominant practices by making them seem like the only practices" that could be imagined, whereas a critical analysis would recognize the constructed nature of the situation and question "how relations of power have shaped discourse practices" (Fairclough 1992: 15).

Wodak and Meyer (2009) identify several branches of CDA and thus encourage people to think of CDA as a *school* of thought that houses diverse approaches. Here, I briefly outline three of the most prominent of these branches. First, the sociocognitive approach put forth by Teun van Dijk (2009) focuses on the relationship between discourse, cognition, and society, taking the position that social structures and discourse structures cannot be directly related to each other as they are mediated through cognition. The sociocognitive approach posits that when interacting with discourse, individuals do not only draw from their own personal experiences, but also upon "collective frames of perception called social representations" (Wodak and Meyer 2009: 26). The sociocognitive approach identifies three forms of social representations relevant to understanding discourse: knowledge, attitudes, and ideologies. Therefore, researchers utilizing the sociocognitive approach would privilege the mental frames of language users. In this way, the sociocognitive approach can be linked to Durkheim's theorizing about the role of collective ideas in both constituting and providing stability for society (Wodak and Meyer 2009).

Next, the dialectical-relational approach, associated with Norman Fairclough (2009), is most concerned with how "dominance, difference, and resistance" are revealed in language and other symbols (Wodak and Meyer 2009: 27). Fairclough (2009) uses the term "semiosis" to refer collectively to language, images, and other symbols that could be counted as discourse, recognizing that language is not the only means by which ideas are transmitted. This approach can be described as grand theory-oriented as its concern is to connect discourse to larger patterns of inequality. Fairclough (2009) refers to this approach as dialectical-relational because in this approach, discourses (including symbols and other images) are conceived of as being dialectically related to other social processes. Fairclough's approach is concerned with social conflict—specifically, the problems faced by oppressed groups. The most recent development in this tradition is to examine "the ways in which and extent to which social changes are changes in discourse" (Fariclough 2012: 452). For example, Fairclough (2012) has examined policy texts to trace social change processes in Romania.

Finally, the discourse-historical approach, associated with Ruth Wodak, is more linguistically oriented than the other approaches (Wodak and Meyer 2009). This approach understands context as historical and is mainly concerned with creating and utilizing existing linguistic tools for understanding specific social problems—namely, those that arise within the political sphere. For example, De Cillia and colleagues (1999) studied the discursive construction of national identities using Austria as a case.

The differences between the three approaches I have just outlined do not constitute practical differences per se. That is to say that researchers using the approaches may use similar or the same methodological techniques, although the historical approach is more linguistically systematic in method in that it involves a more micro-linguistic process. The main differences lie in the underlying theoretical understandings of discourse and the divergent points of focus for each.

This project was informed and inspired, to some extent, by all of these approaches and the general tenets of CDA as a whole—particularly, the idea that discourse is a form of social practice. Related to the sociocognitive approach, I was concerned with Tyson's use of shared cultural ideas and what this use means in terms of how the public interprets Tyson's message. Drawing from Wodak's historical approach, my approach was concerned with a methodical analysis of linguistic devices. That being said, this project probably aligns most (theoretically) with Fairclough's dialectical-relational approach which, compared to other CDA approaches, focuses more wholly on the dialectical relationship between discourse and social action in order to connect everyday discourses with patterns of the larger social structure.

Stibbe (2001) points out that the "power used to oppress animals depends completely on a consenting majority of the human population who, every time it buys animal products, explicitly or implicitly agrees to the way animals are treated" (2001: 147). Hegemonic discourses have the power to manufacture such consent. With the idea of a tolerant "consenting majority" in mind, it is clear that critical discourse analysis is a useful tool for intervening in harm to nonhuman animals.

First, I examined how Tyson represents their *actions*, namely, their actions toward animals and the environment. To analyze Tyson's construction of their actions, I assessed transitivity and verb processes (Halliday 1976; Machin and Mayr 2012). Transitivity concerns whether a verb takes (or can take) a direct object. For example, in the sentence "She hit him," the verb "hit" is transitive as it takes the direct object "him." In the sentence "He ran," the verb "ran" is intransitive because it does not, nor can it, take a direct object. A critical discourse approach to analyzing transitivity goes beyond assessment of grammar to discern who/what is given subject and object statuses. In other words, it allows us to examine who is doing what to whom and thus how action relationships represent power relations (Halliday 1985; Machin and Mayr 2012). The sociological significance of (in)transitivity is that it allows the analyst to discern where agency or responsibility for actions gets "disappeared." Thus, transitivity becomes a tool for analyzing what is *not* being said (Machin and Mayr 2012). In addition, I looked for instances where Tyson represented their actions abstractly where the details of what is actually occurring are obscured. Machin and Mayr note that "when we find such abstractions at the level of social action, we have to ask why and what is being concealed" (2012: 115–116).

My analysis of verb processes utilized Halliday's (1976) typology of verb processes. Halliday (1976) identified six process types: material, mental, behavioral, verbal, relational, and existential.

Material processes are processes that have a material result; statements that present material processes contain an actor (subject), a process (verb), and a goal (object). "He built the house" would be an example of a material process; something concrete has occurred as a result of the action in this statement and there are often beneficiaries of the action represented in material processes. Material processes generally construct an active agent. Machin and Mayr, however, point out that through passive construction of material processes, actors can be "lost" in material processes as in any others (2012: 106). Take, for example, the sentence "The pigs were slaughtered." The agent performing the slaughter is rendered invisible.

Mental processes have to do with "thinking, knowing or understanding" (Machin and Mayr 2012: 107). The use of mental processes can have the effect of garnering empathy for the subject as these types of processes allow a reader or listener to gain insight into the personal thoughts and feelings of the subject. The first part of the sentence "The mother was worried about her sick daughter" is an example of a mental process, which is often constructed in a way that persuades the reader or listener to feel a certain way about the subject and even the object: we likely empathize with both the mother and the child.

Behavioral processes can be described as "a cross between material and mental processes" (Machin and Mayr 2012: 109). With behavioral processes there is some action but that action is experienced personally by the subject. For example, "She tasted the cake" is an example of a behavioral process. Although tasting the cake is an action, it is something psychologically experienced by the subject, but because it involves external action (actions observable by others), it cannot be classified as a mental process.

Verbal processes are actions that denote someone having said something. Verbal processes include a speaker, a recounting of what was said, and often a recipient of what was said, though the recipient can be implied. An example of a verbal process would be "She told the kids to calm down." Examining verbal processes allows us to gauge who in a particular text is given discursive agency or authority.

Related to verbal processes is the use of quoting verbs. Machin and Mayr (2012) discuss how quoting verbs can connote certain meanings about what is being said. For example, compare the following sentences: "The CEO stated that jobs needed to be cut due to constrained budgets" and "The CEO claimed that jobs needed to be cut due to constrained budgets." The first sentence is an example of neutral construction, whereas the second sentence, through the use of the word "claimed," connotes the possibility that what the CEO has said is not true and is open to contestation. By examining which quoting verbs are used in a text, we can see how a text may persuade us to believe or, alternatively, question what is reported as having been said.

Relational processes are processes that have to do with being or having and often describe things/individuals as existing in relation to other things/individuals. For example, "I have tickets to the concert" is a relational process. Other examples include "She is pretty" and "A caterpillar becomes a butterfly." Machin and Mayr point out that the use of relational processes can serve to persuade readers or listeners of discourse to understand certain statements as facts that could actually be classified as opinion. They use the sentence "A lot of people have worries about immigration" as an example of a relational process that appears to present a fact that may just be an opinion or something the writer imagines to be the case (2012: 110).

Finally, existential processes represent happenings in the world and can be described as a cross between of relational and material processes. Like relational processes, existential processes often use the verb "to be" in its various forms. Existential processes, however, contain only one participant that is usually not activated, that is, one participant that is not depicted as having agency. For example, "The sun is shining" is an existential process.

Analysis of process construction allowed me to discern points where Tyson's discourse obscures responsibility for actions and presents opinions as facts. This analysis also allowed me to see where readers might be persuaded to align with Tyson and their ideas, while questioning any opinions that might run counter to Tyson or their practices. That is, it exposes minute processes that constitute hegemonic understandings of corporate conduct where animals are concerned.

I also examined Tyson's use of "nominalization" in the sample pages. Nominalization describes a linguistic device where agency and responsibility for action can be obscured by expressing verb processes as nouns. Machin and Mayr use the concept of globalization/the changed global economy as a clear example: "[the talk] looks at the longer-term picture and examines which countries will emerge in better shape and what should be done to respond to the changed global economy" (2012: 139). Here, the "changed global economy" is presented as a *thing* instead of a *process*. Nothing is said of the who, why, and how of the change. In addition to obscuring agency or responsibility for action, nominalization also functions to obscure those who are affected by the action (Machin and Mayr 2012). Take for example the sentence "The transit strike disrupted travel today." In this instance, it is unclear who the agents are, nor is it clear who is being affected: on the surface, it is travelers who are being affected, but since those who are striking are excluded as agents, we do not even consider their position. The agency of the workers who are striking is obscured. By nominalizing the strike, the wording further serves to represent strikes (and striking workers by default) as problems that disrupt the normal flow of activity. The causes of the strike are backgrounded. So, once processes are presented as things, they can be "counted, described, classified, and qualified" as any other thing may be, which minimizes attention to issues of causality (Machin and Mayr 2012: 142).

Next, I examined the use of presupposition throughout the sample. Presupposition has to do with meanings that are assumed within a text—those pieces of knowledge or understanding that are not within the text but are foundational to the message being conveyed in the discourse. Presuppositions can be relatively benign. Take for example the sentence "The raccoon knocked over my trash can." This sentence about the raccoon presupposes that the reader knows what a raccoon is and also knows what a trash can is. A critical analysis of presupposition allows us to examine what information is presented as given—what is foregrounded and what is backgrounded or silenced (Machin and Mayr 2012). Presupposition is important for connecting Tyson's discourse with larger societal discourses. For example, Tyson states on their website, "We strive to earn consistent and satisfactory profits for our shareholders and to invest in our people, products, and processes." This statement presupposes that pursuit of profit is good and the right thing to do, but more importantly, it presupposes that readers will agree. In short, presuppositions function to present some things as "taken for granted and stable when in fact they may be contestable and ideological" (Machin and Mayr 2012: 137).

Rhetorical tropes are yet another family of linguistic device that I looked for in the discourse of Tyson's website. The specific rhetorical tropes I considered were metaphor, personification/objectification, and metonymy (Machin and Mayr 2012). Lakoff and Johnson note that "our ordinary conceptual system, in terms of which we both think and act, is fundamentally metaphorical in nature" (1980: 3). Given that our conceptual frameworks are steeped in metaphor, it follows that the use of metaphors in various discourses could influence our thinking and the way that we organize our social lives (Machin and Mayr 2012). I should note that discerning the use of metaphor was one of the more difficult techniques used in my analysis as metaphorical understandings of the world are so deeply ingrained. Take for instance the fact that many common phrases used in the United States utilize the metaphor of time as money (e.g., we spend time, we budget time, we lose or run out of time, and we invest time). These sayings are so common and natural to those steeped in US culture that it is difficult to even realize that a metaphor is being used (Lakoff and Johnson 1980). The use of metaphor can allow writers and speakers to avoid specificity in outlining plans or intentions and can also persuade readers to think of things in certain ways. For example, by stating "Our company is interested in building a better future," the speaker invokes the metaphor of physically constructing something which conjures the idea of something concrete and material, although nothing specific is said about how such building will be accomplished or what it even means.

Personification refers to a linguistic device in which human qualities are assigned to things or ideas. Personification can obscure agency. Machin and Mayr use democracy as an example, stating, "Democracy will not stand by while this happens" (2012: 171). In this construction, democracy is cast as the agent; by personifying democracy, the actual agents are concealed and the speaker or writer has also aligned himself/herself with a larger cultural theme that most people are friendly toward.

Metonymy describes an instance where something is referred to by something closely associated with it as opposed to actually calling it by its name. An example of this would be using "Hollywood" to refer to the film industry. A special case of metonymy is synecdoche, where parts of something are used to refer to the whole. Saying "I need another set of eyes on this paper" would be an example of synecdoche where the eyes are used to refer to a person. Although metonymy is similar to metaphor, it is distinct in that metaphors help us to understand things in terms of other things (e.g., understanding social position (class) in terms of spatial metaphors like upper and lower), whereas with metonymy, the primary function is referential, that is, one word/phrase refers directly to another (Lakoff and Johnson 1980). Stibbe (2001) found the use of metonymy to be ubiquitous within the discourse of the meat industry, where, for example, chickens were referred to by their cooking method (e.g., broilers). He notes that the use of metonymy serves to "make the suffering of animals appear unimportant" (Stibbe 2001: 154). In short, rhetorical tropes can be used to abstract processes and agents and to gloss over certain actions (Machin and Mayr 2012).

I scrutinized Tyson's use of narrative in the web pages included in the sample. As Presser notes, stories are constitutive of "who we are and what we intend to do" (2013: 15). Stories also allow the teller to gain social approval by drawing on shared cultural themes (Smith 2005). In considering Tyson's narrative, I use Labov and Waletzky's (1967) definition of a "narrative" as something that is both referential and evaluative—that is, it follows some sequence of events (refers to happenings) and it makes a point (evaluates the happenings in some way). The evaluative piece is what distinguishes narratives from chronicles which simply report happenings with no assessment of the events (White 1980).

Labov (1972) identified six components of a fully formed narrative: *abstract, orientation, complicating action, evaluation, resolution,* and *coda*. The *abstract* offers a short summary of what is to come in the narrative; the *orientation* provides the setting for the narrative and includes details such as the time and place of the happening(s); the *complicating action* is the "body" of the narrative so to speak and is where the sequence of events is recounted; the *evaluation* is where the events are assessed and/or reflected upon; the *resolution* essentially concludes the story; and the *coda* brings the narrative back into the present moment. Using Labov's (1972) framework, I asked which events were storied by Tyson (and which were not), and how these were evaluated. In addition, I attempted to discern points in the narrative where Tyson invoked larger cultural themes.

Finally, I examined the photographs and other images that appeared on the web pages. I analyzed static images. I chose not to analyze videos embedded in the web pages because far less information is available on how to conduct a critical video analysis than on analysis of other forms of discourse. Even considering static images and figures, there are far fewer concrete tools for analysis than that which one can find for critical analyses of language. Indeed, discourse analysts are only recently beginning to include images in their analyses and to theorize how this can be done (see DeLuca 1999; Pace 2005). Photographs and other images are important to discourse analysis as "every image embodies a way of seeing" (Berger 1972: 10) and images both denote and connote (Machin and Mayr 2012). To analyze images featured on the pages in the sample, I critically asked questions about attributes, setting, and salience (Machin and Mayr 2012). Questions regarding the attributes of images included: "Who is pictured in the image?" "How are they posed?" "What are they wearing?" and "What objects are shown in the image and what are the attributes of those objects?" (e.g., Are they clean? Are they colorful?). The setting of images was analyzed by asking questions about the space the image depicts, such as the proximity of objects and people and the use of lighting as in a photograph or the use shading in drawings. Salience is an important factor to consider when critically analyzing images because it is here that we can discern what objects or people are foregrounded or backgrounded in the images—questions such as "What things or people appear the largest in the image?" and "What objects or people are focused on in the image?" By systematically asking these questions about images, we can begin to understand the deep meanings they have to convey.

References
Berger, John. 1972. *Ways of Seeing*. London: Penguin.
De Cillia, Rudolf, Martin Reisigl, and Ruth Wodak. 1999. The Discursive Construction of National Identities. *Discourse and Society* 10: 149–173.
DeLuca, Kevin M. 1999. *Image Politics: The New Rhetoric of Environmental Activism*. New York: Guilford.

Fairclough, Norman. 1992. *Discourse and Social Change*. Cambridge: Polity Press.

Fairclough, Norman. 2009. A Dialectical-Relational Approach to Critical Discourse Analysis in Social Research. In *Methods of Critical Discourse Analysis*, ed. Ruth Wodak and Michael Meyer, 162–186. Los Angeles: Sage.

Fairclough, Norman. 2012. Critical Discourse Analysis. *International Advances in Engineering and Technology* 7: 452–487.

Halliday, Michael A.K. 1976. Types of Process. In *Halliday: System and Function in Language*, ed. Gunther R. Cress, 159–173. Oxford: Oxford University Press.

Halliday, Michael A.K. 1985. *Introduction to Functional Grammar*. London: Edward Arnold.

Josselson, Ruthellen. 2004. The Hermeneutics of Faith and the Hermeneutics of Suspicion. *Narrative Inquiry* 14: 1–28.

Labov, William. 1972. *Sociolingusitic Patterns*. Philadelphia: University of Pennsylvania Press.

Labov, William, and Joshua Waletzky. 1967. Narrative Analysis. In *Essays on the Verbal and Visual Arts*, ed. June Helm, 12–44. Seattle: University of Washington Press.

Lakoff, George, and Mark Johnson. 1980. *Metaphors We Live By*. Chicago: University of Chicago Press.

Machin, David, and Andrea Mayr. 2012. *How to Do Critical Discourse Analysis*. Thousand Oaks: Sage.

Pace, Lesli. 2005. Image Events and PETA's Anti-Fur Campaign. *Women and Language* 28: 33–41.

Presser, Lois. 2013. *Why We Harm*. New Brunswick: Rutgers University Press.

Smith, Philip. 2005. *Why War? The Cultural Logic of Iraq, the Gulf War, and Suez*. Chicago: University of Chicago Press.

Stibbe, Arran. 2001. Language, Power and the Social Construction of Animals. *Society & Animals* 9: 145–161.

van Dijk, Teun A. 1993. Principles of Discourse Analysis. *Discourse and Society* 4: 249–283.

van Dijk, Teun A. 2009. Critical Discourse Studies: A Sociocognitive Approach. In *Methods of Critical Discourse Analysis*, ed. Ruth Wodak and Michael Meyer, 62–86. Los Angeles: Sage.

White, Hayden. 1980. The Value of Narrativity in the Representation of Reality. *Critical Inquiry* 7 (1): 5–27.

Wodak, Ruth, and Michael Meyer. 2009. Critical Discourse Analysis: History, Agenda, Theory, and Methodology. In *Methods of Critical Discourse Analysis*, ed. Ruth Wodak and Michael Meyer, 1–33. Los Angeles: Sage.

Wood, Linda A., and Rolf O. Kroger. 2000. *Doing Discourse Analysis: Methods for Studying Action in Talk and Text*. Thousand Oaks: Sage.

INDEX

Note: Page number followed by 'n' refers to notes.

© The Author(s) 2018
J.L. Schally, *Legitimizing Corporate Harm*, Palgrave Studies in Green
Criminology, https://doi.org/10.1007/978-3-319-67879-5